VOLUME 4: HOW TO PERFORM SKIP-LOT AND CHAIN SAMPLING

SECOND EDITION

The ASQC Basic References in Quality Control: Statistical Techniques
Edward F. Mykytka, Ph.D., Editor

VOLUME 4: HOW TO PERFORM SKIP-LOT AND CHAIN SAMPLING

Second Edition

Kenneth S. Stephens

Volume 4: How to Perform Skip-Lot and Chain Sampling, Second Edition

Kenneth S. Stephens

Library of Congress Cataloging-in-Publication Data

Stephens, Kenneth S.
    How to perform skip-lot and chain sampling / Kenneth S. Stephens.
  —2nd ed.
      p. cm.—(The ASQC basic references in quality control: v. 4)
    Includes bibliographical references and index.
    ISBN 0-87389-325-5
    1. Acceptance sampling. I. Title. II. Series.
TS156.4.S75   1995
658.5'62—dc20
                                               94-30006
                                                 CIP

© 1995 by ASQC

10 9 8 7 6 5 4 3 2 1

ISBN 0-87389-325-5

Set in Times by IPC Publishing & Software Services.
Cover design by Artistic License.
Printed and bound by IPC Publishing Services.

ASQC Mission: To facilitate continuous improvement and increase customer satisfaction by identifying, communicating, and promoting the use of quality principles, concepts, and technologies; and thereby be recognized throughout the world as the leading authority on, and champion for, quality.

For a free copy of the ASQC Quality Press Publications Catalog, including ASQC membership information, call 800-248-1946.

Printed in the United States of America

 Printed on acid-free recycled paper

 ASQC
Quality Press
611 East Wisconsin Avenue
Milwaukee, Wisconsin 53202

The ASQC Basic References in Quality Control: Statistical Techniques is a continuing literature project of ASQC's Statistics Division. Its aim is to survey topics in statistical quality control in a practically usable, "how-to" form in order to provide the quality practitioner with specific, ready-to-use tools for conducting statistical analyses in support of the quality improvement process.

Suggestions regarding subject matter content and format of the booklets are welcome and will be considered for future editions and revisions. Such suggestions should be sent to the series editor(s).

The author wishes to thank the
Series' Review Board, series' editor,
the editorial and production staff
of ASQC Quality Press, and
Thomas M. Kubiak for their assistance
with the current revision.

CONTENTS

FOREWORD TO THE SECOND EDITION

The subject matter of this booklet series continues to be dynamic with contributions by numerous authors and organizations each year. Many of these contributions appear in the journals of ASQC. Hence, periodic revisions of the booklets in the Series is encouraged to keep our readers up-to-date on developments. These may be expressed in the booklets by new areas, expanded areas, and/or as additions to the reference sections to permit the interested reader ease of access to the body of knowledge making up each series' subject. This is certainly the case in the second edition of Volume 4.

Dr. Kenneth S. Stephens, author of the current booklet, recently retired from the United Nations Industrial Development Organization (UNIDO). His last assignment was as Senior Industrial Development Officer, Institutional Infrastructure Branch, Industrial Institutions and Services Division, Department of Industrial Operations, at UNIDO's Vienna, Austria headquarters. He now is active as a consultant on total quality and is a faculty member at Southern TECH, Marietta, Georgia, in the Industrial Engineering Technology Department. He has been associated, earlier, with Western Electric, Rutgers University, LeTourneau College, and Georgia Institute of Technology.

Dr. Stephens worked closely with Harold F. Dodge in the original development of specialty sampling plans. Much of the subject material in this booklet was developed in conjunction with his work with Dodge and the Western Electric Company. Hence, he is able to give the reader a clear picture of both the theory and application of these procedures. His expertise in the area, coupled with substantial industrial and consulting experience as well as academic background, combine to make this booklet one which should be a standard technician's manual as well as a standard classroom and course text and research reference in the area of SkSP and ChSP for many years to come.

The current revision includes an updated and comprehensive reference section that should serve the practitioner and researcher alike for ready access to the large literature associated with this subject. In addition to the procedures, schematics, formulas, derivations, tables, nomographs, and examples for numerous skip-lot and chain sampling plans that were contained in the earlier edition, this second edition includes several new developments in SkSP and ChSP procedures and evaluations. For SkSP, in particular, the *Attribute Skip-Lot Sampling Program*, ANSI/ASQC S1-1987, has been added by way of further explanation, interpretation, and examples. A new appendix has been added to resolve a controversy over an apparent anomaly in several of Dodge's earlier papers.

Edward F. Mykytka
Air Force Institute of Technology/ENS
Wright-Patterson AFB, OH 45433-7765
March, 1994

FOREWORD TO THE FIRST EDITION

The ASQC Basic References in Quality Control: Statistical Techniques is a literature project of the Statistics Division of ASQC. The series' Review Board now consists of Saul Blumenthal, Joseph W. Foster, Alan J. Gross, Gerald J. Hahn, Norman L. Johnson, H. Alan Lasater, Edward A. Sylvestre and Harrison M. Wadsworth, Jr., supplemented (for the current volume, fourth in the series) by Peter R. B. Whittingham.

In Volume I, Number 1 (February 21, 1980) of the *Statistics Division Newsletter*, Philip B. Crosby (President, ASQC) called the comprehension and handling of statistics "... the most basic of needs for all of us ..." He went on to state that "Without numerical information in its most precise form, we cannot complete our responsibility to management and other fellow employees. And without the tools to first comprehend and then explain the analysis we are equally impotent."

This booklet is a companion to Volume 2, How to Perform Continuous Sampling (CSP), both of which were written by Dr. Stephens. In it he describes how to design and use skip-lot and chain sampling plans. These procedures are most often used for the acceptance and rejection of lots where the basic sampling element is the lot itself rather than the acceptance or rejection of units from a continuous process. These plans are especially suited for bulk sampling or for sequences of lots whose order of production is preserved and are extremely useful where the cost of testing is high and minimum sample sizes are imperative.

Dr. Stephens is currently serving with UNIDO on a project with the Mauritius Standards Bureau. He has been associated with the Georgia Institute of Technology and worked closely with Mr. Harold Dodge in the original development of these sampling plans. He has served as a consultant for both industry and governments all over the world. Much of the subject material was developed in conjunction with his work with Mr. Dodge and hence he is able to give the reader a clear picture of both the theory and application of these procedures.

Samuel S. Shapiro
Florida International University
Miami, Florida
April, 1981

Chapter 1

Introduction

In a companion booklet by the author, *How to Perform Continuous Sampling*, Stephens (1995), an introduction to acceptance sampling plans and principles, including types of sampling plans, is presented. That introduction serves, in the main, for the present topics and should be consulted for background information in the study and use of skip-lot and chain sampling. Further to the introduction, those sections on continuous sampling are helpful and necessary for a more complete understanding of skip-lot sampling—being the application of continuous sampling procedures to the entities of lots.

This booklet in the basic reference series describes the principles, procedures, techniques, and applications of skip-lot and chain sampling. While both of these sampling procedures have been around for some time, their potential for widespread application has not been realized. It is the author's hope that recent work on further developments of these procedures along with this booklet will assist in the understanding and applications of these techniques—in addition to further practical developments. An extensive reference section is included to further enhance these possibilities.

Chapter 2

SKIP-LOT SAMPLING PLANS—SkSP

Applying the principles of a Continuous Sampling Plan (CSP-1) to a continuing series of lots or batches of material rather than to individual product units led Dodge (1955b) to develop Skip-Lot Sampling Plans (SkSP). Applications were envisaged to situations where extensive and perhaps costly tests were carried out on characteristics (such as the chemical analysis of the composition of raw materials) of bulk materials or products produced and furnished in successive batches from fairly reliable suppliers (including in-plant movement between production departments).

Applications have included superimposing skip-lot sampling on ordinary lot-by-lot acceptance sampling plans and these ideas have been formalized by Dodge and Perry (1971); Perry (1970, 1973a, 1973b); and more recently by ANSI/ASQC Standard S1-1987 (1987).

2.1 BASIC PARAMETERS, OPERATION, INTERPRETATION, AND APPLICATION

The basic parameters of SkSP-1 are those of CSP-1, namely the clearing interval, i, and the sampling fraction, f. The operation is also similar to CSP-1. Two procedures have been proposed by Dodge (1955b) for SkSP-1. One applies to reprocessing and correcting nonconforming lots or replacing nonconforming lots with conforming ones. This procedure applies when nonconforming lots are readily corrected or replaced as would be possible if the inspection was being used between departments within a factory or on certain receiving inspection situations where the supplier is relatively close and accessible for such corrections or replacements. The second procedure applies to rejecting and removing nonconforming lots without requiring that they be corrected or replaced. It is particularly applicable to situations in which rejected lots (found nonconforming during inspection, whether during the clearing interval or during sampling) are returned to a supplier for disposition—with no expectation that the lots will be corrected and returned.

A schematic for these procedures is given in Figure 2.1. This schematic assumes that skip-lot sampling has been chosen for the application and that the parameters, i and f, have been selected to achieve a given AOQL type protection on the average outgoing quality of *lots*, by quality characteristic.

<table>
<tr><td>At the outset for each characteristic under consideration, test every lot consecutively as purchased or supplied.</td></tr>
</table>

<table>
<tr><td>Continue such testing until " i " lots in succession are found to be conforming for one or more quality characteristic(s).</td></tr>
</table>

<table>
<tr><td>Begin sampling of lots on a frequency of " f " fraction of lots and test each sampled lot for the quality characteristic(s) qualified above.</td></tr>
</table>

<table>
<tr><td>As long as no lots are found to be nonconforming for the quality characteristic(s) being tested, continue to sample and test only the fraction " f " of the lots.</td><td>When a nonconforming lot is found, revert again to testing every consecutive lot for the quality characteristic(s) found to be nonconforming.</td></tr>
</table>

Figure 2.1. Operation Schematic for SkSP-1

For Procedure 1, submit all lots found nonconforming for reprocessing, correction or replacement, and resubmission with respect to the quality characteristic(s) that caused the nonconformance.

For Procedure 2, replace "i" in the schematic of Figure 2.1 with "i + 1" and remove (e.g., return to the source of supply) all nonconforming lots.

When referring to specific CSP-1 plans (upon which SkSP plans are based) with parameters **i** and **f** and related **AOQ** characteristics, Dodge (1955b) mentions, "It can be shown that **i** should be increased by one in CSP-1 plans when defective units are removed but not replaced." Dodge's Procedure A1 (the replacement case) uses **i**, whereas his Procedure A2 (the nonreplacement case) uses (**i + 1**). When referring to the formulas that apply when defective (nonconforming) units* found are corrected or **replaced** with good units (known as the replacement case), Dodge (1943) has noted that the substitution of **i** by (**i − 1**) yields the **formulas** for the **nonreplacement** assumption. Since these developments and the related statements by Dodge in the cited papers (of 1943 and 1955), there have been occasions when these statements were thought to be contradictory and in some cases incorrectly applied. This subject is considered in greater detail in Appendix B, where it is shown that both statements are correct and the applicable conditions for each case are clearly elucidated.

Specific plans by AOQL, i, and f may be selected from the CSP-1 curves of Figure 2.2. However, it is envisaged that SkSP-1 would not be used with very small AOQLs, large values of i, or small values of f. Hence, a reduced graph of CSP-1 curves is given in Figure 2.3, especially for use in selecting SkSP-1 plans. It is from Dodge (1955b). Other graphs such as Figures 2.16 and 2.17 of Stephens (1995) can also be used. Some practical selections from these curves are listed in Figure 2.4.

Dodge (1955b) has summarized the comparison of terms applicable to CSP-1 and SkSP-1. This is reproduced in Figure 2.5 and serves to illustrate the interpretations that apply to SkSP-1.

---

* For definitions of terms such as defective, defect, nonconformity, and nonconforming unit, see ANSI/ASQC Standard A2-1987 (1987) as well as *Glossary and Tables for Statistical Quality Control*, ASQC (1983). All of these terms are used in various places in this booklet. The reader is advised, in any instance, to substitute whichever term applies to the situations for which the principles and techniques are being used.

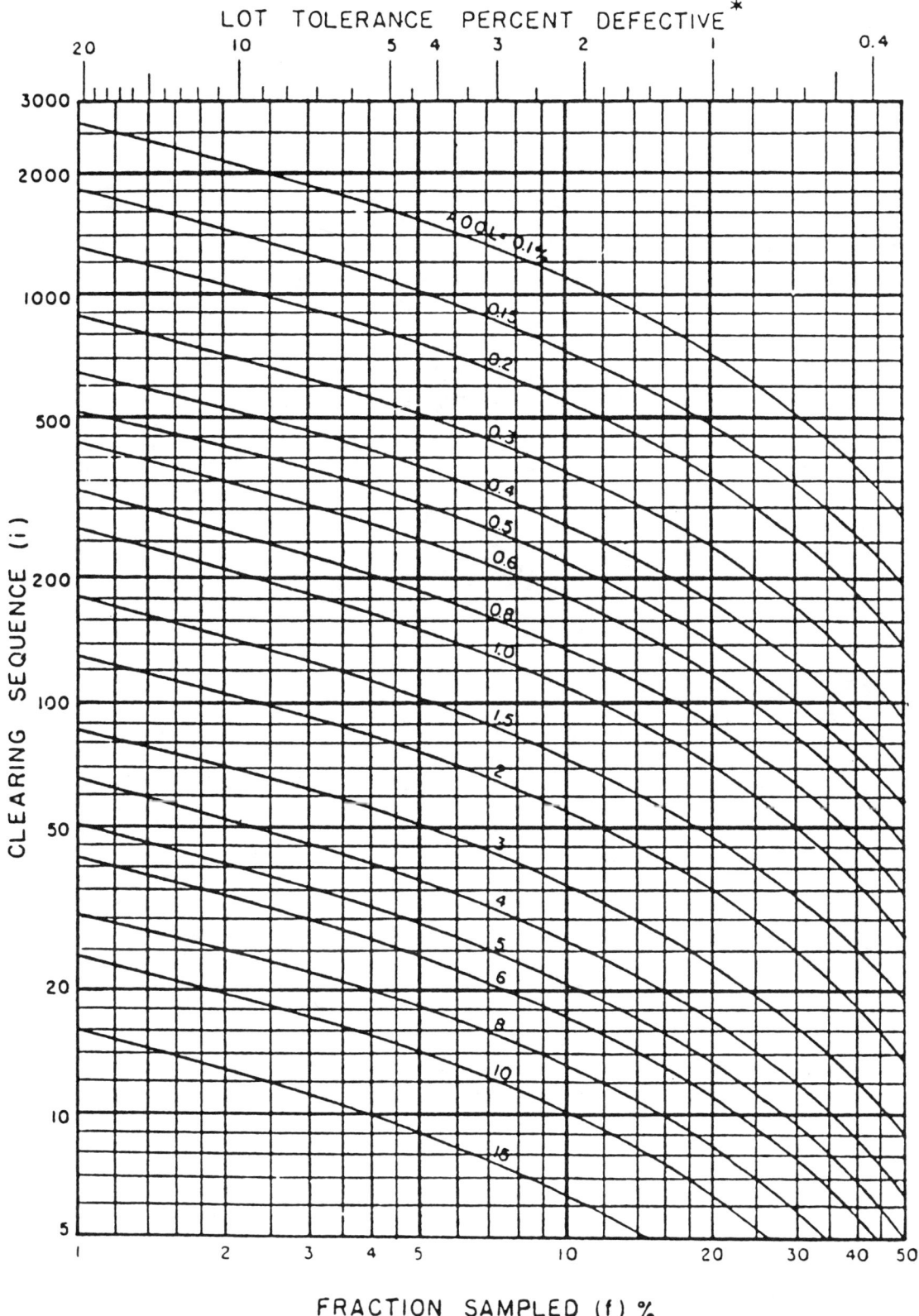

*For a run of 1,000 units at a consumer's risk of 10%

Figure 2.2. Nomograph for Selecting CSP-1 Sampling Plans

5

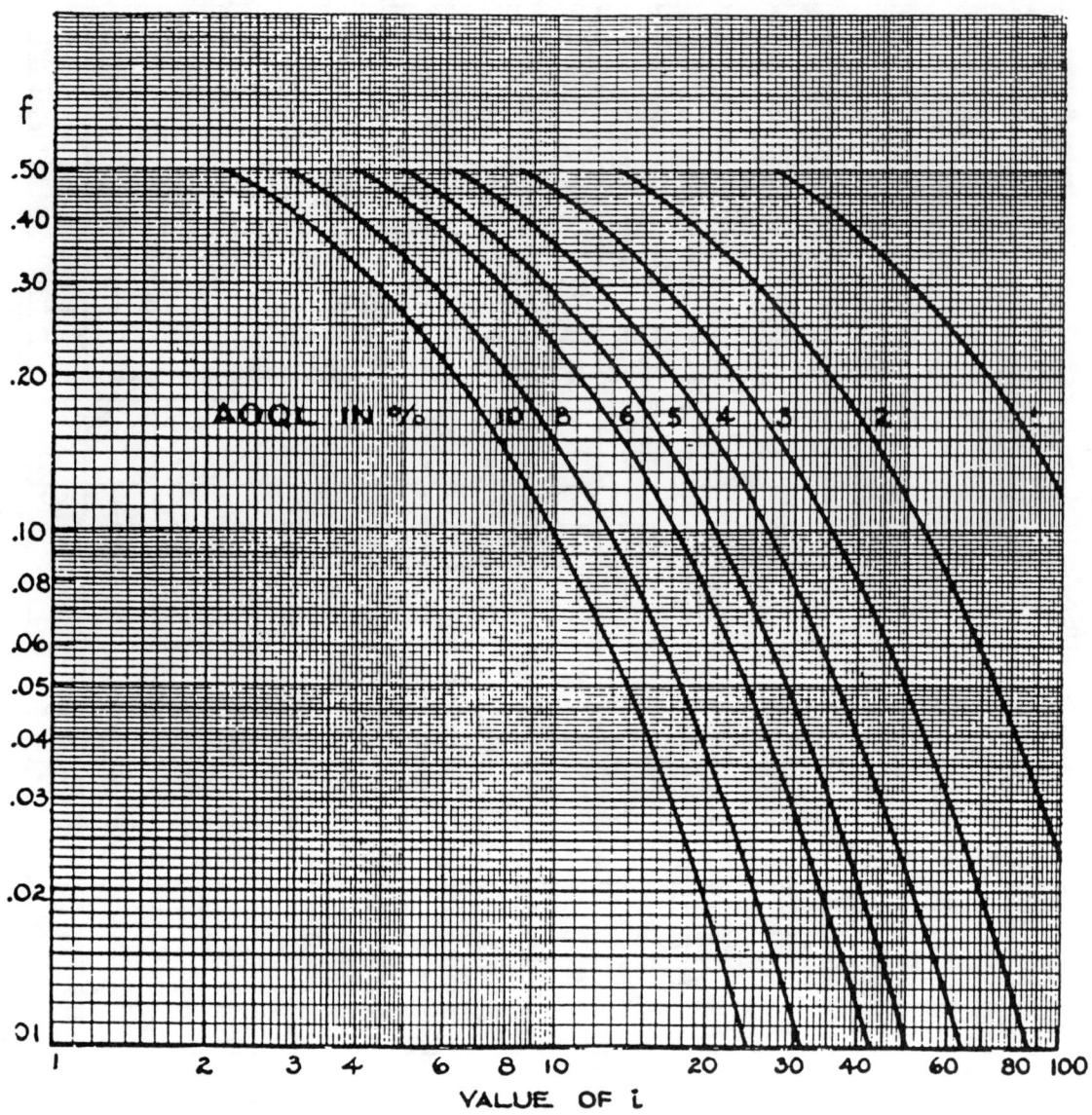

Figure 2.3. Nomograph for Determining Values of f and i for a Given AOQL for CSP-1, Small Values of i

AOQL	f	Procedure 1 i	Procedure 2 i +1
1%	1/2	27	28
2%	1/2	14	15
	1/3	23	24
3%	1/2	9	10
	1/3	15	16
4%	1/2	7	8
	1/3	11	12
	1/4	14	15

Figure 2.4. Practical Parameter Selection for SkSP-1

As the basic parameters and operation of SkSP-1 are similar to those of CSP-1, so also are the operating characteristics with modification in interpretation to reflect the differences outlined in Figure 2.5. Hence, the formulas for CSP-1 given in section 2.1 of Volume 2, Stephens (1995), apply.

SkSP-1 is generally considered applicable to situations of continuing supply from the same source normally expected to be of the same satisfactory quality. It is usually associated with lots for which a single determination or analysis is made to ascertain the acceptability of the lots. More than one quality characteristic may be checked, with each treated separately at the prescribed AOQL or treated collectively against the prescribed AOQL of the plan. It often relates to lots for which bulk sampling or composite sampling is used to obtain a representative sample upon which the analysis is made.

CSP-1 (product units)	**SkSP-1** (lots of a raw material)
series of units	series of lots (or batches)
inspect a unit	make lab analysis of a sample of material
nonconforming unit (a unit that fails to meet the applicable specification requirement)	nonconforming lot (a lot where the sample fails to meet the applicable specification requirement)
units in succession found clear of nonconformities	lots in succession found conforming
incoming % nonconforming: % of incoming units that are nonconforming	incoming % nonconforming: % of incoming lots that are nonconforming
meaning of 2% AOQL: an average of not more than 2% of accepted units will be nonconforming for the characteristics under consideration.	meaning of 2% AOQL: an average of not more than 2% of accepted lots will be nonconforming for the characteristics under consideration.

Figure 2.5. Comparison of Terms for CSP-1 and SkSP-1

2.2 EXTENSIONS OF SKIP-LOT SAMPLING PLANS AND APPLICATIONS

Dodge and Perry (1971) and Perry (1973a) formalize the application of skip-lot sampling to the situation in which each lot to be inspected is sampled according to some attribute (with possible extension to variable) lot-inspection plan. They label the plan as SkSP-2. Perry (1973b) further extends the procedures to two-level skip-lot sampling plans.

2.2.1 SkSP-2

The specifications of SkSP-1 have not precluded the use of a lot-by-lot acceptance sampling plan for assessing the conformance of each inspected lot. However, the operating characteristics for such a combination were not explicit. As indicated above, this was made explicit in the papers by Dodge and Perry (1971) and Perry (1973a) and labeled SkSP-2.

In addition to the parameters of SkSP-1, namely the i and f of the underlying CSP-1 plan, SkSP-2 includes a given lot-inspection plan by the method of attributes (such as single sampling, double sampling, multiple sampling, sequential sampling, etc.) called the "reference sampling plan". It is thus more of a system of sampling with the underlying CSP-1 plan dictating how the reference sampling plan is to be used.

An operation schematic for SkSP-2 is shown in Figure 2.6. Evaluation of the operating characteristics of SkSP-2 combines an analysis of the underlying CSP-1 plan with parameters i and f and the operating characteristics of the reference sampling plan. Let P denote the probability of acceptance of a lot according to the reference plan and Pa(f,i) the corresponding probability of acceptance for the SkSP-2 plan.

Dodge and Perry (1971) give the following results for SkSP-2. The phase during which every lot is sampled and inspected according to the reference plan is referred to, by Dodge and Perry, as "normal inspection." With the advent of ANSI/ASQC S1-1987 (1987) and the direct application of skip-lot sampling with ANSI/ASQC Z1.4 providing the reference plan(s) (see section 2.3), this phase will now be referred to as "qualification inspection" so as not to confuse this with **normal** inspection of the *normal,*

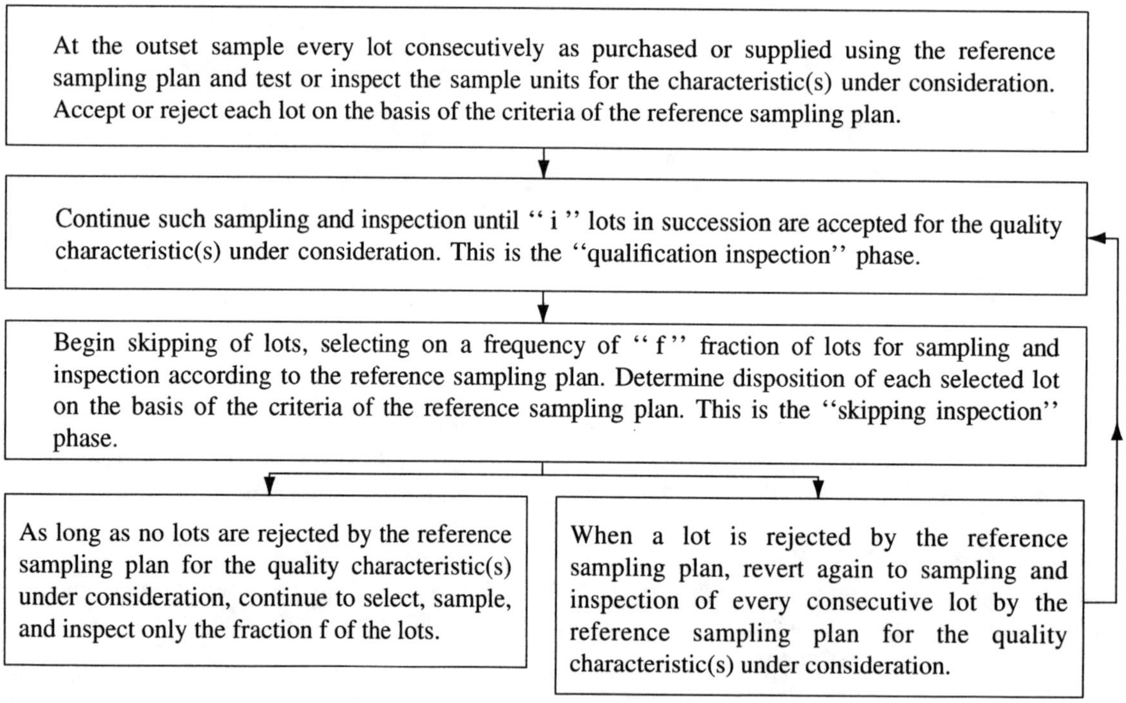

Figure 2.6. Operation Schematic for SkSP-2

8

reduced, and tightened inspection levels of the Z1.4 plans. The phase during which lots are skipped and selected for sampling and inspection with fraction f is referred to (by Dodge and Perry **and** in this booklet) as "skipping inspection".

1. The average or expected number of lots inspected during qualification inspection, U, where

$$U = (1 - P^i)/[P^i(1 - P)] \tag{2-1}$$

2. The average or expected number of lots inspected during skipping inspection, V, where

$$V = 1/[f(1-P)] \tag{2-2}$$

3. The average fraction of the total lots that are inspected, F, where

$$F = (U + fV)/(U + V) = f/[f + (1-f)P^i] \tag{2-3}$$

4. The average fraction of total lots accepted, Pa(f,i) for the SkSP-2 plan, where

$$Pa(f,i) = FP + (1-F), \tag{2-4}$$

since lots that are inspected (with probability F) have probability P of being accepted, while skipped lots (not inspected with probability, 1-F) have unit probability of acceptance. Hence,

$$Pa(f,i) = [fP + (1-f)P^i]/[f + (1-f)P^i] \tag{2-5}$$

Another of the common measures of evaluation is that of AOQ and its associated AOQL. As explained for SkSP-1, this requires interpretation in terms of lots instead of product units. However, with consideration of a reference plan applied to the sampling and inspection of product units in a lot under SkSP-2, the more customary measure of AOQL (on product units) can be examined though not always with ease. To facilitate distinction, the following definitions are used.

1. AOQ_1, the average fraction (or percentage) of outgoing product units that are defective, and its upper limit, $AOQL_1$.

2. AOQ_2, the average fraction (or percentage) of outgoing lots that are nonconforming (i.e., are rejected or would be rejected by the reference sampling plan), and its upper limit, $AOQL_2$.

When the reference sampling plan is a single sampling plan, Perry (1970) developed the expression for $AOQL_1 = Y/n$, which is similar to the Dodge-Romig (1959) expression for a single sampling plan, AOQL = y/n − y/N. Perry (1970) also developed a table of Y factors for various values of f, i, and n/N as shown in Appendix A. It enables selection of single sampling reference plans for ratios of n/N of 0 (large N), .10, .20, and .50; for i values of 4, 6, 8, and 10; for f values of 2/3, 1/2, 1/3, 1/4, and 1/5; and acceptance numbers, c, of 1 to 10 in increments of 1. As an example, for the single sampling plan of n = 50, c = 1, and N = 500, Figure 2.7 lists the Y values and corresponding AOQLs (in %) associated with the tabled values of i and for f = 1/2 and 1/5.

For other reference sampling plans, such as double or multiple sampling plans, no simple way of expressing $AOQL_1$ has been developed.

AOQ_2, on the other hand, in terms of lots, the basic entity of the underlying CPS-1 plan used with SkSP-2, is easily expressed as,

$$AOQ_2 = (1-P)(1-F), \tag{2-6}$$

and as Perry (1973a) notes, can be further expressed as,

$$AOQ_2 = Pa(f,i) - P, \text{ see } Pa(f,i) \text{ in equation (2-4) above.} \tag{2-7}$$

Thus, as observed by Perry (1973a), AOQ_2 is the increase over the probability of acceptance of the reference sampling plan, P, representing the nonconforming lots that are skipped (and hence accepted) by SkSP-2. He further develops a table of $AOQL_2$ values for a practical series of f and i values for SkSP-2.

9

			i			
			4	6	8	10
f	1/2	Y	.8910	.7749	.7609	.7572
		$AOQL_1$	1.78	1.55	1.52	1.51
	1/5	Y	.9626	.8441	.7832	.7619
		$AOQL_1$	1.93	1.69	1.57	1.52

Figure 2.7. Selected Values of Y and $AOQL_1$ for an Example Reference Single Sampling Plan for SkSP-2

This is shown in Figure 2.8. The reader should note, however, that the parameters of Figure 2.4 and hence of Figures 2.2 and 2.3 apply equally to $AOQL_2$ determinations of SkSP-2 plans, as does Figure 2.8 to SkSP-1. They represent different forms of displaying the AOQL results for convenience in selecting parameters of suitable plans.

Beyond the above common operating characteristics for SkSP-2 plans, Perry (1973a) also develops the Average Sample Number (ASN) and the Average Run Length (ARL) to detect (via lot rejection) an abrupt deterioration of quality.

For ASN, the following simple expressions apply:

$$ASN \ (SkSP) = ASN \ (R) \cdot F, \qquad (2\text{-}8)$$

where ASN (SkSP) is the average sample number for the skip-lot plan, ASN (R) is the average sample number for the reference sampling plan, and F is as given by equation (2-3) above.

For a single sampling reference plan for which ASN (R) = n,
$$ASN \ (SkSP) = n \cdot F. \qquad (2\text{-}9)$$

i	f						
	1/5	1/4	1/3	2/5	1/2	3/5	2/3
4	.148	.126	.089	.081	.060	.044	.034
6	.105	.089	.069	.057	.042	.030	.024
8	.081	.069	.053	.044	.032	.023	.018
10	.066	.056	.043	.035	.026	.019	.015
12	.056	.047	.037	.030	.022	.016	.013
14	.048	.041	.031	.026	.019	.014	.011

Figure 2.8. Values of $AOQL_2$ for f and i Values

For good quality processes, savings in sampling can be significant and this serves to emphasize the potential use of SkSP-2 as a system for achieving reduced inspection (as discussed in greater detail in section 2.2.3).

For ARL, the following definitions and expressions apply (see also Stephens [1966], page 77)
ARL-1 is the average number of lots to the occurrence of a rejection after the process quality shifts abruptly from p_0 to p_1 ($p_0 < p_1$) with the count beginning on the first lot after the shift (counting both skipped and inspected lots).

$$ARL\text{-}1 = \frac{1}{1-P_1} + \frac{(1-f)[P_0^i + f(1 - P_0) \sum_{j=0}^{i-1} P_0^j P_1^{i-j}]}{f(1 - P_1)[f + (1 - f) P_0^i]} \tag{2-10}$$

where P_0 and P_1 are the probabilities of acceptance for the reference sampling plan for process quality at p_0 and p_1 respectively.

ARL-2 is the average number of lots to the occurrence of a rejection following a rejection.

$$ARL\text{-}2 = \frac{1}{1 - P_a(f,i)} \tag{2-11}$$

Perry (1973a) indicates the relationship, ARL-1 ≥ ARL-2, develops curves for ARL-1 for a number of SkSP-2 plans with a single sampling reference plan, and concludes that changes in f are more influential on ARL-1 than are changes in i. Thus, if fast response to abrupt changes in quality are desired, larger values of f should be chosen.

Perry's (1973a) paper should be consulted for additional details on ASN and ARL.

Additional assistance, particularly in the selection of a single sampling reference plan for SkSP-2, is provided by Dodge and Perry (1971) in the form of a table of operating ratios, OR, for various f and i values for SkSP-2 plans and for lot-by-lot single sampling plans. The operating ratio is defined as the ratio of the fraction defective for which there is a probability of acceptance of 0.10 (i.e., p_{10}) and the fraction defective for which there is a probability of acceptance of 0.95 (i.e., p_{95}); hence, OR = p_{10}/p_{95}. Together with np_{95}, the OR values form a basis for deriving sampling plans with desired operating characteristics, under conditions where Poisson probabilities can be used as suitable approximations to binomial probabilities. The Dodge-Perry table is given in Figure 2.9.

The use of this table for deriving an SkSP-2 plan with desired properties is illustrated by the following example.

A skip-lot plan is to be instituted. Desired operating characteristics include a satisfactory producer's quality level of 2% defective with a 95% probability of acceptance (AQL = 2.0%), and a limiting quality level of 8% defective with a 10% probability of acceptance (LQL = 8%). These values can be expressed as p_{95} = 0.02 and p_{10} = 0.08. Hence, the desired OR = p_{10}/p_{95} = 0.08/0.02 = 4.0. Entering the table in Figure 2.9 with this OR value locates the SkSP-2 plan with f = 1/2 and i = 4 having an OR = 4.063.

For this plan, the single sampling reference plan will have an acceptance number, c, of 3 and a sample size obtained from solution of the equation, np_{95} = 1.645. Hence,

n = $1.645/p_{95}$ = 1.645/0.02 = 82.25 ≈ 83.

The SkSP-2 plan is then: n = 83, c = 3 as the reference sampling plan with f = 1/2 and i = 4 as the skipping parameters.

Additional information from Figure 2.9 for this plan is that it is matched reasonably well with a regular (no skipping) single sampling plan with c = 4 and n = 100. This latter value is obtained by dividing the sample size of the reference sampling plan by the value in the last column (to

11

Single Sampling Plan			Skip-Lot Plan, SkSP-2				Ratio of SkSP-2 Sample Size to Matched Single Sampling Plan
c	OR	$np_{.95}$	(f,i)	c	OR	$np_{.95}$	Sample Size
2	6.509	0.818	(1/5,8)	1	6.505	0.598	.731
3	4.890	1.365	(1/5,14)	2	4.883	1.090	.799
4	4.057	1.970	(1/2,4)	3	4.063	1.645	.835
5	3.549	2.613	(1/2,6)	4	3.522	2.270	.869
			(1/2,8)	4	3.574	2.237	.856
			(1/5,8)	3	3.561	1.876	.718
6	3.206	3.285	(1/2,10)	5	3.207	2.892	.880
			(1/4,8)	4	3.191	2.505	.762
7	2.957	3.981	(1/2,12)	6	2.951	3.569	.894
			(1/4,10)	5	2.930	3.166	.795
			(1/5,14)	5	2.963	3.130	.789
			(1/5,6)	4	2.982	2.681	.673
8	2.768	4.695	(2/3,4)	7	2.757	4.270	.909
			(2/3,6)	7	2.777	4.238	.902
			(1/2,14)	7	2.759	4.266	.908
			(1/4,14)	6	2.778	3.791	.807
			(1/5,8)	5	2.782	3.334	.709
9	2.618	5.425	(2/3,6)	8	2.611	4.977	.917
			(2/3,8)	8	2.627	4.947	.912
			(1/2,4)	7	2.627	4.482	.826
			(1/3,10)	7	2.597	4.533	.835
			(1/5,10)	6	2.629	4.007	.738
			(1/5,4)	5	2.594	3.578	.659
10	2.497	6.169	(2/3,8)	9	2.493	5.698	.924
			(2/3,10)	9	2.505	5.670	.919
			(1/3,14)	8	2.507	5.184	.840
			(1/5,12)	7	2.506	4.698	.761
			(1/5,6)	6	2.499	4.215	.683

Figure 2.9. Operating Ratios for Single Sampling Plans and SkSP-2 Plans with Single Sampling Reference Plans

the right) of Figure 2.9, e.g., 83/.83 or 100 (or $n = np_{.95}/p_{.95} = 1.970/.02 \approx 99$ using the third column of Figure 2.9).

2.2.2 TWO-LEVEL SKIP-LOT SAMPLING PLANS

With CSP-1 as the underlying plan providing the skipping parameters for SkSP-1 and SkSP-2, it is somewhat natural to examine the possible application of the multi-level continuous sampling plans to skip-lot inspection. This has been done by Perry (1970 and 1973b).

Three procedures are developed and labeled 2L.1, 2L.2, and 2L.3. Their operation and principal operating characteristics are as follows.

2.2.2.1 PLAN 2L.1

1. Start with qualification inspection, using the reference sampling plan for every lot.
2. When i consecutive lots are accepted on qualification inspection, switch to skipping inspection at rate f_1.
3. During skipping inspection at rate f_1:
   (a) When i consecutively inspected lots are accepted, switch to skipping inspection at rate f_2.
   (b) When a lot is rejected, switch to qualification inspection for every lot.
4. During skipping inspection at rate f_2, when a lot is rejected, switch to skipping inspection at rate f_1.

$$Pa^{2L.1}(f_1,f_2,i) = \{f_2P^i+f_1[P(1-P^i)-P^i(1-P^{i+1})] + (f_1-f_2)P^{2i}\}$$
$$/\{f_2[P^i+f_1(1-P^i)^2]+(f_1-f_2)P^{2i}\}. \tag{2-12}$$

where f_1, f_2, and i are the parameters of the procedure and P is the probability of acceptance of the reference plan for process quality, p.

This plan is based on the MLP procedures of Lieberman and Solomon (1955), which are similar to the CSP-M procedures of section 2.2.3 Volume 2, Stephens (1995), with $K = 2$ and without the rule of 4 (not feasible for the small values of i envisaged for skip-lot applications). However, the only restriction of the skipping rates f_1 and f_2 is that $f_1>f_2$. Exponential reduction in the f's is not assumed.

2.2.2.2. PLAN 2L.2

1. Start with qualification inspection, using the reference plan for every lot.
2. When i consecutive lots are accepted on the qualification inspection level, switch to skipping inspection at rate f_1.
3. When i consecutively inspected lots are accepted on skipping inspection at rate f_1, switch to skipping inspection at rate f_2.
4. When a lot is rejected on either skipping inspection level, switch to qualification inspection for every lot.

$$Pa^{2L.2}(f_1,f_2,i) = \{f_2[P^i+f_1(P-P^i)] + (f_1-f_2)P^{2i}\}/\{f_2[P^i+f_1(1-P^i)] + (f_1-f_2)P^{2i}\},$$
$$\text{with } f_1,f_2,i, \text{ and P as for 2L.1.} \tag{2-13}$$

This procedure is based on the tightened multi-level continuous sampling plans of Derman, Littauer, and Solomon (1957) that are similar to CSP-T of section 2.2.4 in Volume 2, Stephens (1995). However, as for 2L.1, the only restriction on f_1 and f_2 is that $f_1>f_2$. Geometric reduction of the f's is not assumed.

2.2.2.3 PLAN 2L.3

1. Start with qualification inspection using the reference plan for every lot.
2. Note when i consecutive lots are accepted on qualification inspection.
   (a) If this occurs on the ith lot (i.e., first i lots accepted), switch to skipping inspection at rate f_2.
   (b) If this occurs later than the ith lot, switch to skipping inspection at rate f_1.
3. When a lot is rejected on either skipping inspection level, switch to qualification inspection for every lot.
$$Pa^{2L.3}(f_1,f_2,i) = Pa^{2L.2}(f_1,f_2,i)$$

13

This procedure is novel, is seen to have identical operating characteristics of 2L.2, and may have appeal for easier administration. It has, in fact, been used in the structure of the skip-lot sampling program of ANSI/ASQC S1-1987 (1987), as will be discussed in section 2.3.

For further details on these procedures consult Perry (1970 and 1973b) and Brugger (1975). For information on a cost model for SkSP consult Hsu (1977 and 1980), as well as Callejas (1976), and Bennett and Callejas (1980). See Reimann (1982) for a counter discussion. For a consideration of inspection error applied to a skip-lot plan see Carr (1982). Other skip-lot plans and evaluations are given by Cox (1982), Heldt (1981), Jafri (1988), Lenz and Rendtel (1984), Osborne (1990–91), Parker and Kessler (1981), Phelps (1982), Reetz (1984), and Stine (1974).

2.2.3 SkSP-2 APPLIED TO REDUCED INSPECTION

A significant application for SkSP-2 is its use as a procedure for reduced inspection in a sampling scheme (such as MIL-STD-105E or ANSI/ASQC Z1.4) involving switching between various inspection levels depending on indications of the quality of incoming lots. The following quotation from Dodge and Perry (1971) pertains.

> "The skip-lot plan SkSP-2 provides a framework that is particularly well-suited as a basis of 'reduced inspection' for such structures as the composite scheme of Normal and Reduced Inspection used in MIL-STD-105D* and like systems of sampling inspection of lots.
>
> "For example, MIL-STD-105D uses a complete set of 'reduced inspection' sampling plans to be substituted for the set of Normal Inspection sampling plans when certain criteria are met, notably 'when the preceding 10 lots or batches have been on Normal Inspection and none has been rejected on original inspection'. Basically the plan for reduced inspection calls for a sample of reduced size from **each** lot about 40% as large as the sample for Normal Inspection together with an appropriate acceptance number, so as to give an OC curve for Reduced Inspection that is suitably looser than the OC curve for Normal Inspection. Instead of using a reduced sample size plan for each lot for Reduced Inspection, there are definite merits in using the Normal Inspection under normal conditions and just skipping the inspection of an appropriate fraction, f, of the lots for Reduced inspection. In this way, just the regular sampling plans for Normal Inspection are used— no additional special reduced sample size plans are needed.
>
> "Specifically, for MIL-STD-105D, a reduced inspection plan comparable to the one given in 105D would be to use SkSP-2, with the inspection plan given in 105D for Normal Inspection as the 'reference plan' together with f = .40 (inspecting only 2 out of each 5 lots) and an i = 10. The reduction in inspection would be about 60% when quality is good (process fraction defective, p, is small), as it is in 105D. Other choices of f and i are of course possible."

With the publication of the ANSI/ASQC Standard S1-1987 (1987) a standardized scheme for Skip-Lot Sampling Inspection (of the SkSP-2 type) has been established. The scheme is intended to be used in conjunction with ANSI/ASQC Z1.4-1993 (1993) which is more or less equivalent to MIL-STD-105E (1989). This standard provides the lot-by-lot phase of the scheme. As discussed in the previous quotation from Dodge and Perry, the S1 Standard is essentially a scheme to substitute for the reduced inspection procedure of ANSI/ASQC Z1.4, although as a part of its operation it also allows reduced inspection during the Skip-Lot Inspection qualification period. This, of course, could be altered by the Responsible Authority stipulating that the reduced inspection part of Z1.4 is not to be used. Some details of the scheme are discussed in section 2.3.

---

* MIL-STD-105D is now replaced by MIL-STD-105E (1989), as shown in the reference section. However, the arguments presented are still valid and have, in fact, been formalized in a Skip-Lot Sampling Standard, ANSI/ASQC Standard S1-1987 (1987). See section 2.3 for details.

14

2.3 ANSI/ASQC STANDARD S1-1987

Unlike the CSP procedures in Volume 2, Stephens (1995), and its numerous references, except for the S1 Standard, there are no coordinated tables or extensive tabulations indexed by quality measures for Skip-Lot Sampling Inspection. In general, plans must be devised individually, with some convenience provided by the brief tabulations of Figures 2.4 and 2.8 and the formulas for evaluations. But even the referenced figures provide no immediate assistance in the selection of a reference sampling plan for SkSP-2 applications.

In this connection the reader should note the significance of the resource contained in the earlier quotation from the Dodge and Perry (1971) paper—it is used as the basis for the lot-by-lot phase of the ANSI/ASQC S1 scheme. The single, double, and multiple sampling plans of MIL-STD-105E are well documented, including their operating characteristics. These are potential reference sampling plans* for SkSP-2, as are other tabulations of lot-by-lot acceptance sampling plans—quite apart from the use of SkSP-2 as a system of reduced inspection for a multi-level inspection system such as ANSI/ASQC Z1.4— as provided by ANSI/ASQC Standard S1.

> For example: For inspection of lots of size 91 to 150 and a desired AQL of 2.5%, the single sampling plan for normal inspection from MIL-STD-105E, using code letter F, is $n = 20$, $c = 1$. As a reference plan for SkSP-2 for various values of f and i, the OC curves and ASN curves are presented by Perry (1973a) and illustrate the relationships.

The S1 Standard is also not a collection or tabulation of f and i values to be used with the skip-lot procedure together with a reference lot-by-lot sampling plan to achieve various AOQL or AQL protection. In fact, for all indices of AQL (used in S1) the f and i values are essentially the same. They are part of a standardized procedure (similar to Perry's 2L.3 plan but with expansion from two levels to four levels), with "i" incorporated in the operating procedure (with values of 20, 10, and 4 at various stages) and "f" values fixed at 1/2, 1/3, 1/4, and 1/5 at various levels.

The attainment of different AQL protection is actually achieved by (1) selection of the reference sampling plan (scheme) under Z1.4, and (2) variable criteria incorporated in tabulations of required cumulative and individual lot sampling results.

Hence, the Attribute Skip-Lot Sampling Program of ANSI/ASQC Standard S1 is quite unique, and as mentioned in the Standard, "should be distinguished from Dodge's skip-lot plans."

2.3.1 BASIC OPERATION AND EVALUATION OF S1

The interested reader and practitioner should obtain a copy of the ANSI/ASQC Standard S1-1987 as well as a copy of ANSI/ASQC Z1.4. Reference should also be to the paper by Liebesman and Saperstein (1983). This paper describes the basic procedure and presents various analyses and the rationale for structuring the switching criteria. While this is still valid, it should be noted that the procedures finally adopted in S1 differ somewhat from those given by Liebesman and Saperstein. In particular, the initial qualification for skip-lot inspection at the various levels of 1/2, 1/3, and 1/4 was modified, the table for individual lot acceptance numbers for skip-lot qualification was changed, and the disqualification from Skip-Lot Interrupt was changed.

An operation schematic for the Attribute Skip-Lot Sampling Program is shown in Figure 2.10. It is expressed in the form of a flow chart in the S1 Standard as Figure 1 together with criteria for three conditions stipulated in the flow chart.

In addition to the individual and cumulative lot criteria in the operation schematic of Figure 2.10, the S1 Standard includes other qualifications and approvals for Skip-Lot Inspection.

---

* MIL-STD-105E is intended to be used as a system of sampling plans, see Stephens and Larson (1967), but may be used as a resource or reservoir of well-documented plans for individual selection and application. See the example that follows.

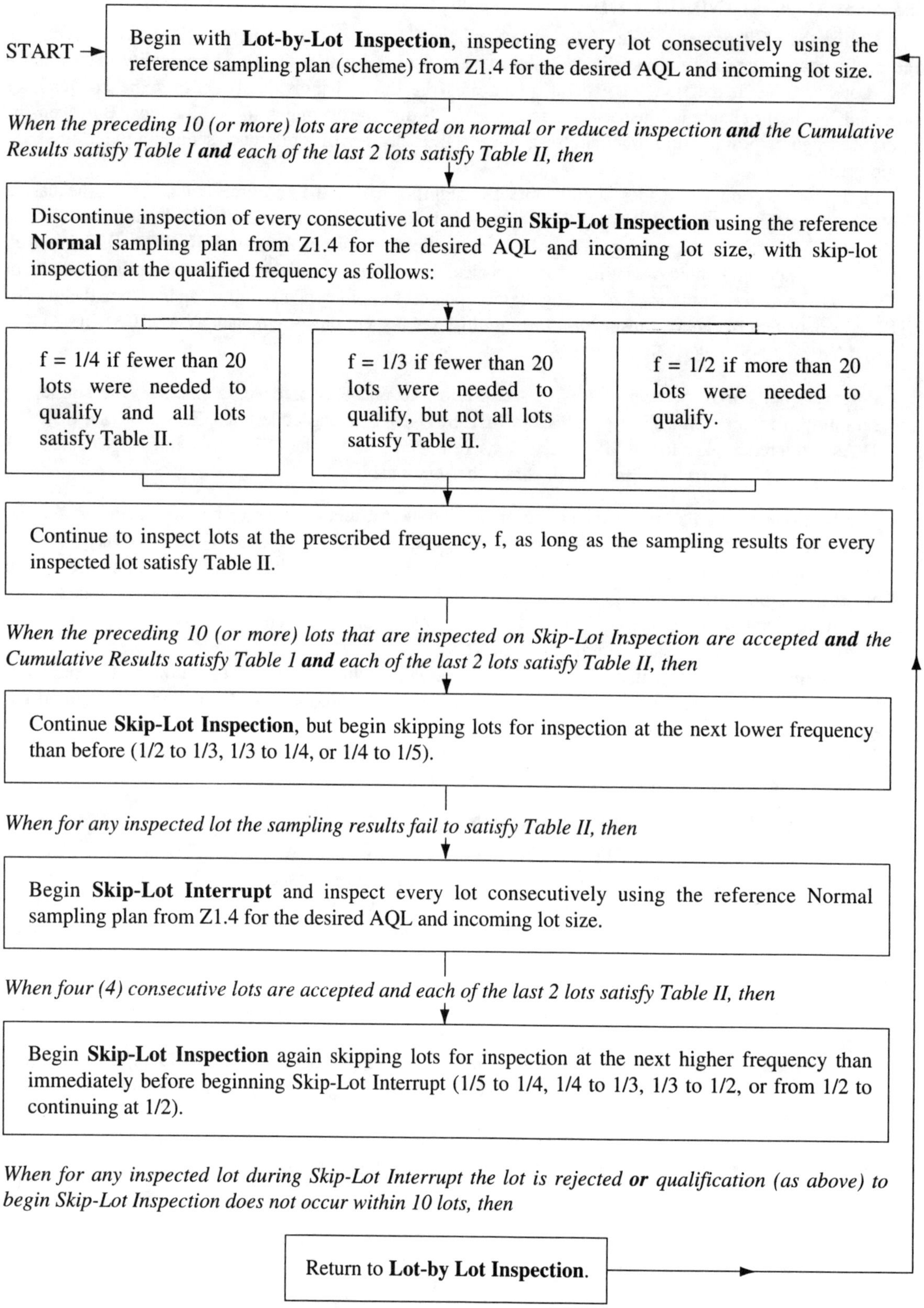

START → | Begin with **Lot-by-Lot Inspection**, inspecting every lot consecutively using the reference sampling plan (scheme) from Z1.4 for the desired AQL and incoming lot size.

*When the preceding 10 (or more) lots are accepted on normal or reduced inspection **and** the Cumulative Results satisfy Table I **and** each of the last 2 lots satisfy Table II, then*

Discontinue inspection of every consecutive lot and begin **Skip-Lot Inspection** using the reference **Normal** sampling plan from Z1.4 for the desired AQL and incoming lot size, with skip-lot inspection at the qualified frequency as follows:

| $f = 1/4$ if fewer than 20 lots were needed to qualify and all lots satisfy Table II. | $f = 1/3$ if fewer than 20 lots were needed to qualify, but not all lots satisfy Table II. | $f = 1/2$ if more than 20 lots were needed to qualify. |

Continue to inspect lots at the prescribed frequency, f, as long as the sampling results for every inspected lot satisfy Table II.

*When the preceding 10 (or more) lots that are inspected on Skip-Lot Inspection are accepted **and** the Cumulative Results satisfy Table 1 **and** each of the last 2 lots satisfy Table II, then*

Continue **Skip-Lot Inspection**, but begin skipping lots for inspection at the next lower frequency than before (1/2 to 1/3, 1/3 to 1/4, or 1/4 to 1/5).

When for any inspected lot the sampling results fail to satisfy Table II, then

Begin **Skip-Lot Interrupt** and inspect every lot consecutively using the reference Normal sampling plan from Z1.4 for the desired AQL and incoming lot size.

When four (4) consecutive lots are accepted and each of the last 2 lots satisfy Table II, then

Begin **Skip-Lot Inspection** again skipping lots for inspection at the next higher frequency than immediately before beginning Skip-Lot Interrupt (1/5 to 1/4, 1/4 to 1/3, 1/3 to 1/2, or from 1/2 to continuing at 1/2).

*When for any inspected lot during Skip-Lot Interrupt the lot is rejected **or** qualification (as above) to begin Skip-Lot Inspection does not occur within 10 lots, then*

Return to **Lot-by Lot Inspection**.

Figure 2.10. Operation Schematic for ANSI/ASQC Standard S1

16

To enter Skip-Lot Inspection initially and each time after disqualification, it is required to meet Supplier and Product Qualifications as follows:

Supplier Qualification is stated as,

"The Supplier Shall:

A. Have implemented and maintained a documented system for controlling product quality and design changes. It is assumed that such a system includes inspection by the supplier of every lot produced and recording of inspection results.
B. Have instituted a system which is capable of detecting and correcting shifts in quality levels, and monitoring process changes which may adversely affect quality. The supplier's personnel responsible for the application of the system shall exhibit a clear understanding of the applicable standards, systems, and procedures to be followed.
C. Not have experienced an organization change that might adversely affect quality."

Product Qualification is stated as,

"The Product shall have met all of the following requirements:

A. Be of stable design.
B. Have been manufactured on an essentially continuous basis for a period mutually agreed to by the supplier and responsible authority. If no period is specified, the period shall be six months. Whenever production is held pending sample approval, only the time period after approval and resumption of production shall be included.

NOTE: Essentially continuous production is considered a stabilizing factor of the manufacturing or assembly process.

C. Have been on normal or reduced inspection or a combination of normal and reduced inspection at general inspection levels I, II, or III (see ANSI Z1.4) during the qualification period. A product that has been on tightened inspection during the qualifying period is ineligible for skip-lot inspection.
D. Have been maintained at the AQL quality level or better (see ANSI Z1.4) for a period of stability mutually agreed to by both the supplier and responsible authority. If no period is specified, the period shall be six months."

Additionally, in order to begin Skip-Lot Inspection and/or to continue Skip-Lot Inspection at a lower frequency, application of the S1 Standard requires the approval of the **Responsible Authority**. While in the Skip-Lot Inspection or the Skip-Lot Interrupt states, the S1 Standard also requires return to the Lot-by-Lot Inspection state if: (1) there is no production during the agreed to period; (2) the supplier violates the requirements of Supplier or Product Qualification (as above) or additional factors apply, such as a documented complaint, the supplier has knowledge as to which lots are to be inspected, the supplier fails to maintain a quality assurance system that includes inspection of each lot produced and records of the results; and (3) the Responsible Authority desires to return to Lot-by-Lot Inspection.

The S1 Standard makes provision for the application of Skip-Lot Inspection as a **substitute** for reduced inspection. Given in ANNEX C of the Standard are factors for deciding between Skip-Lot Inspection (that uses the Normal Inspection plan during the skip-lot states) and the regular Reduced Inspection of ANSI/ASQC Z1.4. (See also the discussion in section 2.2.3.)

For all of the above and more, the S1 Standard should be consulted. Tables I and II of the standard are reproduced in Appendix C by permission of ASQC.

Chapter 3

Chain Sampling—ChSP

Another of the specialized types of sampling procedures developed by Dodge (1955a) is that of Chain Sampling Plans (ChSP). The development and introduction were contemporary with skip-lot sampling plans - both were presented by Dodge at the same conference in February 1954.* The plans were published in consecutive issues of *Industrial Quality Control*, ChSP in January, 1955, and SkSP in February, 1955. Bell Telephone Laboratory internal memoranda on the procedures had been written even earlier.

The special purpose of ChSP is to achieve a more favorable operating characteristic than that of single sampling, especially with an acceptance number, c, of zero, for relatively small samples motivated by situations of destructive or costly tests, satisfying conditions suitable to the use of cumulative results of samples from several lots in making an acceptance decision on the current lot. Individual lots and samples are considered as links in a chain; hence, the name, chain sampling. Considerable extensions and variations have been developed from the basic procedure.

3.1 BASIC PARAMETERS, OPERATION, EVALUATION, AND APPLICATION

The first chain sampling plan by Dodge (1955a), designated ChSP-1, employs two parameters, n and i, where n is the sample size or number of units selected at random from each lot for inspection for the specified requirement(s) and i is the number of preceding sample results that, if clear of defects (defective units), allows acceptance of the current lot when one defective unit is found in its sample. Otherwise, acceptance of the current lot is permitted when no defects (defective units) are found in its sample. The acceptance criteria are fixed at zero or one as noted above. The lot size is not designated as an explicit parameter, being of insignificant consequence under binomial or Poisson sampling. An operation schematic for ChSP-1 is shown in Figure 3.1.

---

* Annual Middle Atlantic Regional Conference, ASQC, Baltimore, Maryland, February 5, 1954.

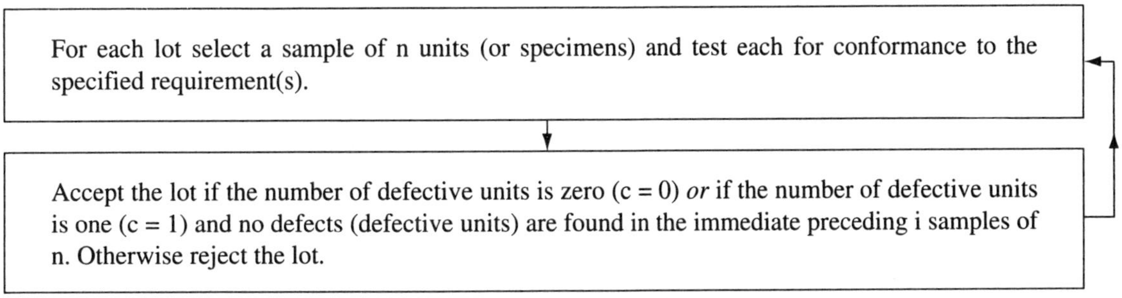

For each lot select a sample of n units (or specimens) and test each for conformance to the specified requirement(s).

Accept the lot if the number of defective units is zero (c = 0) *or* if the number of defective units is one (c = 1) and no defects (defective units) are found in the immediate preceding i samples of n. Otherwise reject the lot.

Figure 3.1. Operation Schematic for ChSP-1

Evaluation of ChSP-1 is by means of the probability of acceptance, Pa, the operating characteristic that is readily computed from results derived by Dodge (1955a), viz,

$$\text{Pa} = P_{0,n} + P_{1,n}(P_{0,n})^i, \text{ where} \tag{3-1}$$

$P_{0,n}$ = probability of finding no defects (defective units) in a sample of n units from product of fraction defective, p.

$P_{1,n}$ = probability of finding one defect (defective unit) in such a sample.

The formula may be used for drawing OC curves for values of n and i for a series of p values. An example of such OC curves for several values of i is shown in Figure 3.2. It is from from Dodge (1955a). Binomial probabilities are used. The single sampling plan with n = 10 and c = 0 is also illustrated and clearly shows the relationship between chain sampling and single sampling.

These curves have a somewhat different meaning than the usual probability of acceptance curves for the ordinary types of lot-by-lot sampling plans, due to the cumulative nature of the acceptance criterion. They show, for given n and i, the percentage of lots expected to be accepted for a given value of product quality, p, and the evaluation assumes that p remains fixed over the accumulation period. For ordinary lot-by-lot sampling, decisions are based on information from each lot and are not dependent on data or the quality level of previous lots.

The curves of Figure 3.2 also serve to illustrate the ChSP effect that is succinctly stated by Dodge (1955a) as follows: "Curves for individual ChSP-1 plans can be compared with the OC curves for basic c = 0 plans of single sampling. It is seen that adding the provision for using cumulative results for i preceding samples has the same effect on the characteristic curve as taking a second sample. It increases the chances of acceptance in the region of principal interest - where the product percent defective is very small. Since in addition it calls for rejection provided only that two defects are fairly close together, it modifies the basically undesirable features of the c = 0 single sampling plans".

Additional OC curves for ChSP-1 plans are presented by Clark (1960). He argues the case for larger sample sizes and presents OC curves for various sample sizes up to n = 100. He further illustrates an application of ChSP-1 for the sampling inspection of a device for a failure characteristic.

Further evaluations of plans of the ChSP-1 type have been carried out by Zwickl (1965) and Soundararajan (1978a and 1978b). These relate to the determination of AOQL for ChSP-1 and tabulations of ChSP-1 plans for AQL/LQL and AQL/AOQL criteria. Under conditions suitable to the use of Poisson probabilities (in its own right or as approximations to binomial and hypergeometric probabilities), the following expression can be used to determine the AOQL of ChSP-1 plans.

$$\text{AOQL} = k/n - k/N. \tag{3-2}$$

Values of k for the parameter i for ChSP-1 are given in Figure 3.3.

A nomograph for selecting parameters n and i for ChSP-1 and AOQL, based on the above result, is given in Figure 3.4.

Based on the use of binomial probabilities, the following expressions have been derived for the parameters n and i of ChSP-1 having an OC curve passing approximately through the two points, (p,Pa) : $(p_1, 0.95)$ and $(p_2, 0.10)$, with p_1 as AQL and p_2 as LQL (see section 1.3 of Stephens [1995]).

$$n = \log(0.10)/\log(1-p_1) \text{ and} \tag{3-3}$$

$$i = \{[\log[0.95-(1-p_1)^n]-\log(np_1)]/[n \log(1-p_1)]\} - (n-1)/n \tag{3-4}$$

Based on these results a table of some practical ChSP-1 plans indexed by a series of AQL and LQL values is given in Figure 3.5. Values of i have been rounded to the nearest integer. The table is from Soundararajan (1978b).

An additional expression for determining a ChSP-1 plan of fixed sample size, n, and desired OC curve points, (p, Pa):$(p_1, 1-\alpha)$, (p_2, β), which for given values of p_1 and p_2 minimizes $(\alpha + \beta)$ is as follows:

i is the integer nearest to:

20

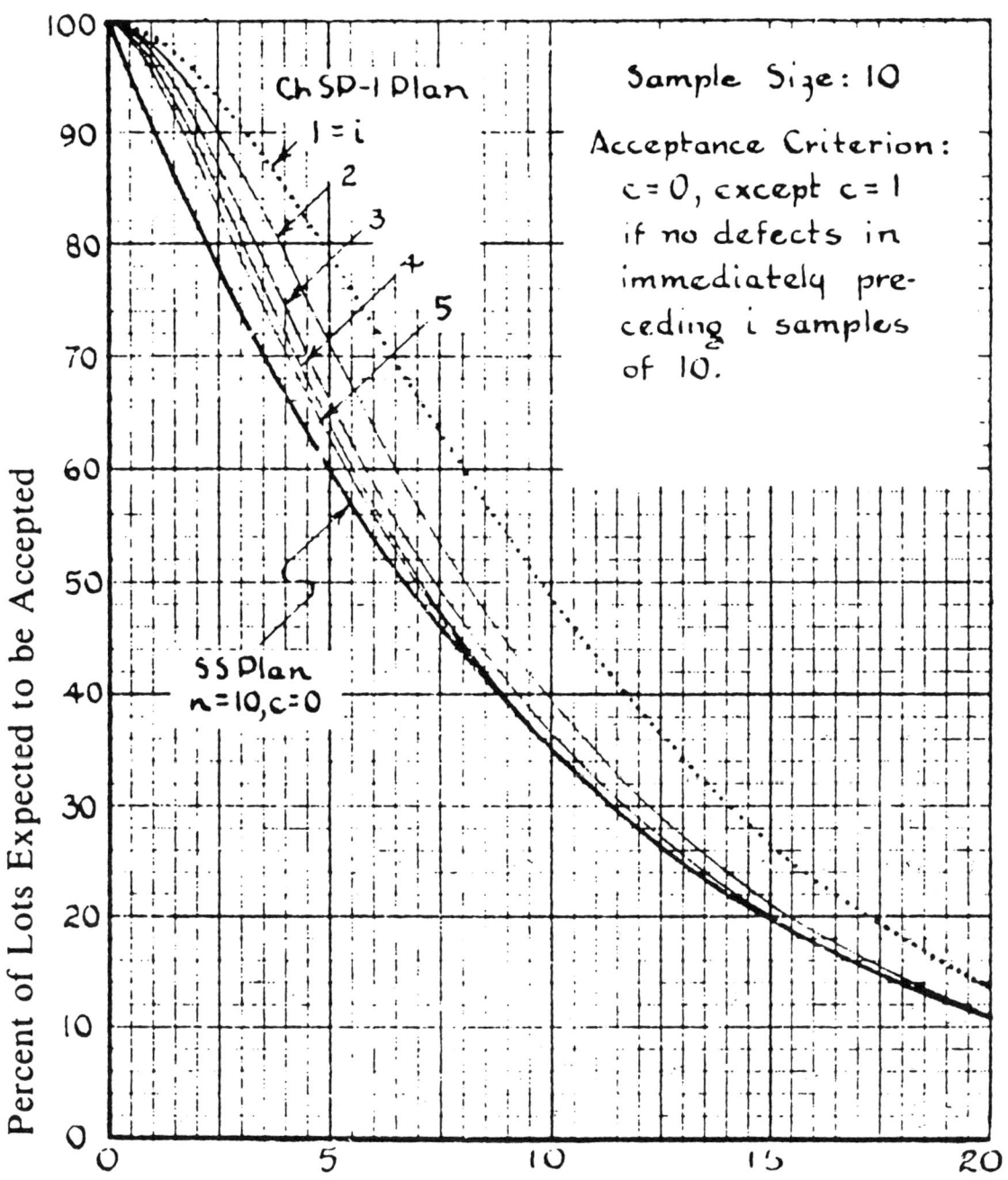

Figure 3.2. OC Curves for ChSP-1 Plans with Values of i from 1 to 5; n = 10

i	k
1	.503
2	.419
3	.389
4	.376
5	.372
$\vdots$	$\vdots$
∞ *	.368

* Corresponds to single sampling with c = 0

Figure 3.3. AOQL Factors for ChSP-1

$$\frac{1}{2} + \frac{1}{n} + \frac{\log(p_2/p_1)}{n\log[(1-p_1)/(1-p_2)]} + \frac{\log[(1-p_2)^n - 1]/[(1-p_1)^n - 1]}{n\log[(1-p_1)/(1-p_2)]}$$

For example: for $n = 20$, $p_1 = .005$, $p_2 = .05$; $i = 4$.

Chain sampling plans are designed for applications satisfying the following conditions:

1. Lots should be a series of continuous supply, preferably in order of production.
2. Normally, lots are expected to be of essentially the same satisfactory quality.
3. The consumer has confidence in the integrity of the producer.

Chain sampling plans are particularly suited to situations satisfying the above conditions and involving destructive, complex, and/or costly tests for which a small sample is desired yielding good quality discrimination. They are additionally useful for achieving better discrimination (than single and double sampling plans) in situations associated with assuring tighter quality levels. Zwickl (1965) illustrates an application to transistors requiring an AOQL = 0.2% for which the ChSP-1 plan with $n = 195$, $i = 2$ involves less Average Total Inspection (ATI) and less relative total inspection costs than the single sampling plan, $n = 180$, $c = 0$ and others.

3.2 EXTENSIONS OF CHAIN SAMPLING PLANS

Early work on extensions of chain sampling plans included plans designated ChSP-2 and ChSP-3 in an unpublished memorandum by Dodge (1958). While these were presented at the 13th Midwest QC Conference of ASQC in November 1958, they were never published - perhaps owing to the administrative complexities of several parameters, normal and reduced inspection levels, and provision for a second sample on reduced inspection.

Frishman (1960) presents extended chain sampling plans designated ChSP-4 and ChSP-4A (perhaps contemplating publication of designations 2 and 3 by Dodge). His plans evolved from an application in the sampling inspection of torpedoes for Naval Ordnance, Frishman (1954), as a check on the control of the production process and test equipment (including 100% inspection). Features of the plans include a basic acceptance number greater than 0, an option for forward or backward cumulation of results for an acceptance-rejection decision on the current lot, and provision for rejecting a lot on the basis of the results of a single sample (ChSP-4A).

Some variations of chain sampling for which cumulative results are used in the sentencing of lots have also been developed by Anscombe, Godwin, and Plackett (1947); Page (1955); Hill, Horsnell, and

22

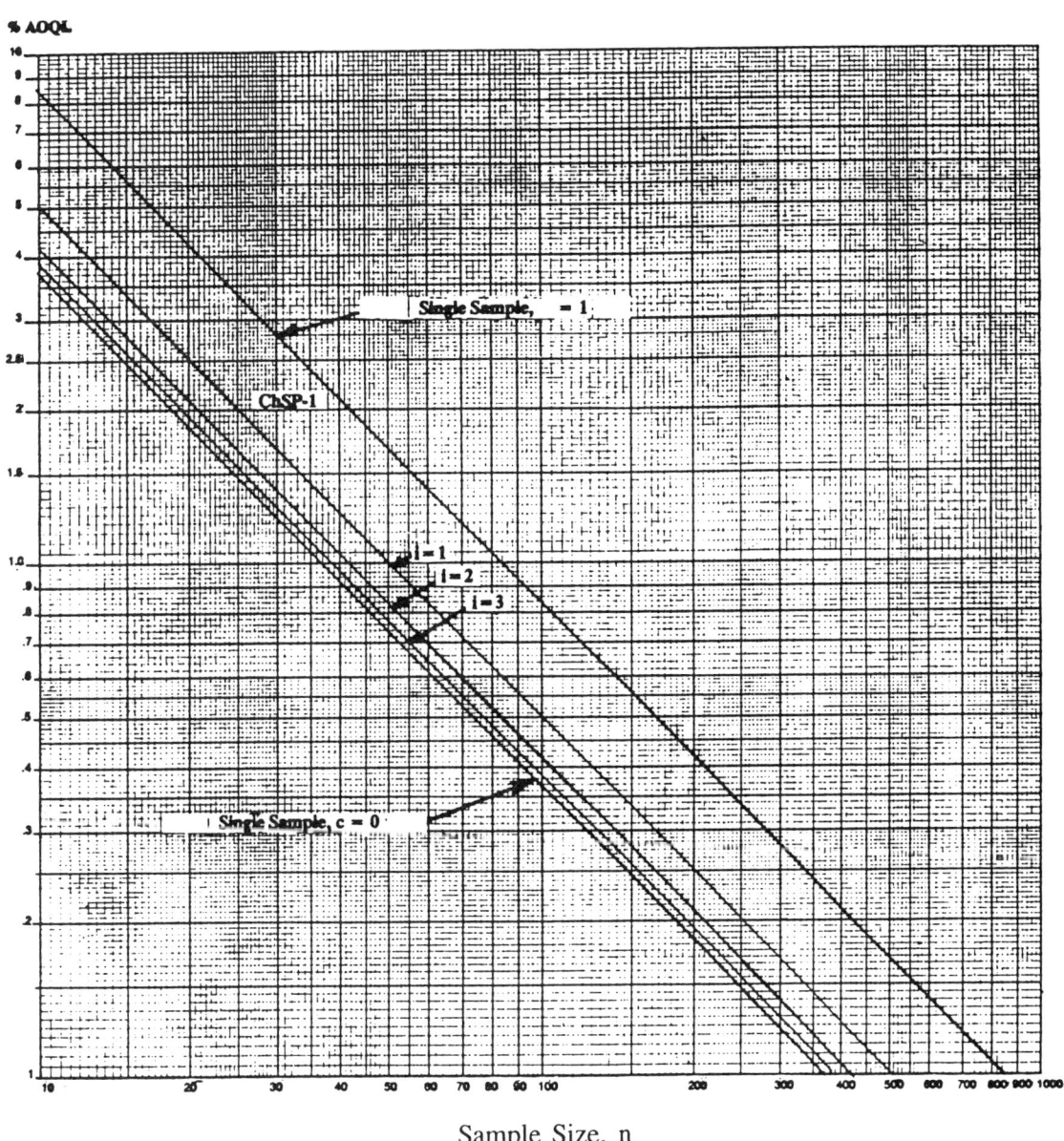

Sample Size, n

Figure 3.4. Nomograph for Selection of Parameters n and i for ChSP-1 and Desired AOQL

23

LQL in percent	Sample size	0.10	0.15	0.25	0.40	0.65	1.00	1.50	2.50	4.00	6.50
					AQL in percent						
1.0	228	2									
1.5	152	4	1								
2.0	114	7	2								
2.5	91		3	1							
3.0	76		4	2							
3.5	65			2							
4.0	57			3	1						
4.5	50			4	2						
5.0	45			5	2						
5.5	41			7	3						
6.0	38			9	3						
6.5	35				4	1					
7.0	32				5	1					
7.5	30				6	1					
8.0	28				7	2					
8.5	26					2					
9.0	25					2					
9.5	23					3					
10.0	22					3	1				
11.0	20					4	2	1			
12.0	18					5	2	1			
13.0	17					5	2	1	1		
14.0	16						2	1	1		
15.0	15						3	1	1		
16.0	14						3	2	1	1	
17.0	13						4	2	1	1	
18.0	12						5	2	1	1	
19.0	11						6	3	1	1	
20.0	11						6	3	1	1	
21.0	10						7	3	1	1	
22.0	10						7	3	1	1	
23.0	9							4	1	1	
24.0	9							4	1	1	
25.0	8							5	2	1	
30.0	7							7	2	1	
35.0	6								2	1	
40.0	5								4	2	1
50.0	4								7	3	1
60.0	3									6	2
70.0	2									8	4

Figure 3.5. Values of n and i Indexed by AQL and LQL for ChSP-1

Warner (1959); Ewan and Kemp (1960); Kemp (1962); Beattie (1962); Cone and Dodge (1964); and Wortham and Moog (1970). Further details on these procedures can be obtained from the literature cited.

Further extensions to a general family of chain sampling inspection plans have been developed by Dodge and Stephens and published in numerous technical reports, conference papers, and journal articles (see the Reference section for Dodge and Stephens [1964, 1965, and 1966], Stephens [1966, 1979, 1980, and 1995], and Stephens and Dodge [1965a, 1965b, 1966a, 1966b, 1966c, 1967, 1974, 1976a, and 1976b]). Some details on these plans are presented in the following subsections as well as in sections 3.3 and 3.4.

3.2.1 ChSP-4 and ChSP-4A

The parameters for these plans as given by Frishman (1960) are as follows, with some symbols changed for consistency with the other sections.

n - sample size drawn from each lot.

C_1 - acceptance number of the first stage.

k - number of lots in the first and second stages combined, over which cumulation of sampling results takes place. Corresponds to i + 1 in ChSP-1.

C_2 - acceptance number for k lots combined, i.e., the cumulative results criterion.

R_1 - rejection number of the first stage (ChSP-4A).

An operation schematic for these plans is shown in Figure 3.6.

Evaluation is by means of the OC curve for which Frishman (1960) gives the following general formulas for Pa, the probability of acceptance.

$$\text{For ChSP-4, } Pa = \Pr(d \leq C_1 \mid n,p) + \Pr(D \leq C_2 \mid C_1 < d < C_2, kn, p) \qquad (3\text{-}6)$$

$$\text{For ChSP-4A, } Pa = \Pr(d \leq C_1 \mid n,p) + \Pr(D \leq C_2 \mid C_1 < d < R_1, kn, p) \qquad (3\text{-}7)$$

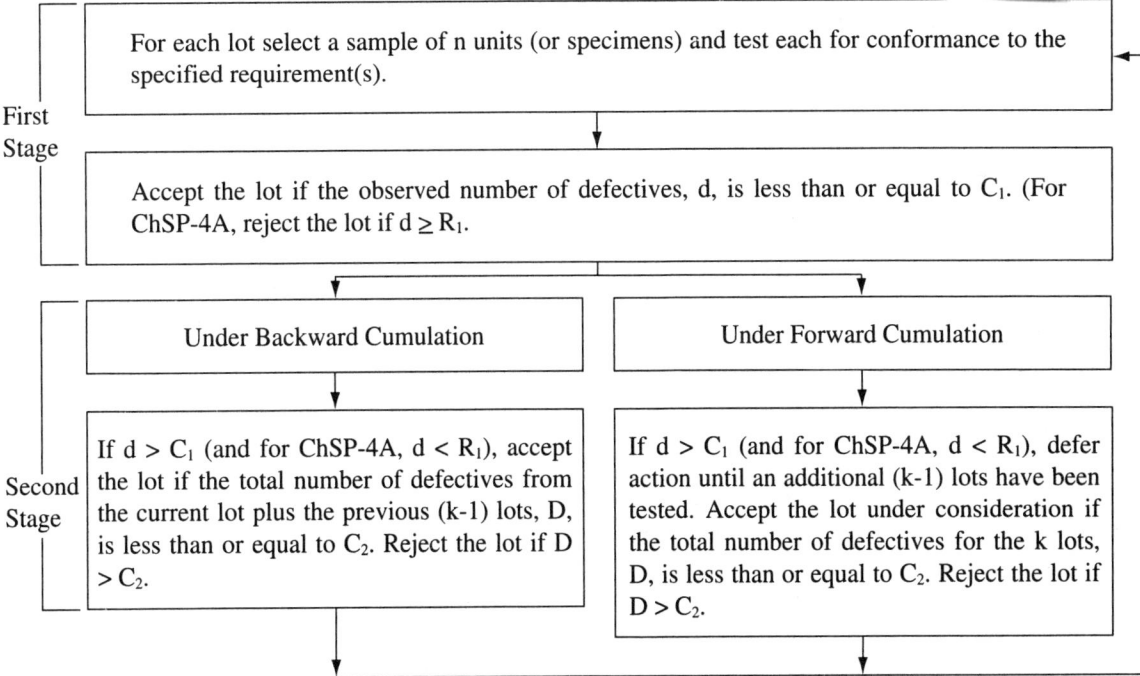

Figure 3.6. Operation Schematic for ChSP-4 and ChSP-4A

25

OC curves for several plans are given in Frishman (1960) to illustrate the effects of (1) changes in the sample size - being tighter plans with greater discrimination for larger sample sizes, (2) changes in the size of second stage, i.e., k - being somewhat tighter plans having larger k values, (3) changes in the rejection number R_1 - being slightly tighter plans in the region of good quality for smaller values of R_1 (not necessarily a desirable feature), and (4) adding the second stage to the first stage - being the "chain sampling effect" of a higher probability of acceptance in the region of principal interest where the product percent defective is very small. The first stage is an ordinary single sampling plan with n and C_1. The second stage is the chain sampling feature using cumulative results.

Tables of these plans are not available. They must be tailor-made for specific applications as was done in Frishman (1960).

3.2.2 GENERAL FAMILY OF CHAIN SAMPLING INSPECTION PLANS

In the series of publications by Dodge and Stephens (see Reference section), a whole family of chain sampling plans have been developed and described. These are designated ChSP-C_1,C_2 and ChSP (n_1,n_2)-C_1,C_2 for the cases of equal sample size and different sample sizes, respectively, in the generalized family of two-stage chain sampling inspection plans. Parameters for these plans are k_1,k_2,C_1,C_2, and n or n_1,n_2. Alternative designations based on the parameter values are $(k_1,k_2; C_1,C_2)$ for given n and $(n_1,n_2; k_1,k_2; C_1,C_2)$ for explicit description of the plans with different sample sizes in the two stages. The parameters are defined as follows:

k_1 - the maximum number of samples over which the cumulation of defectives (defects) takes place in the first stage of the procedure.

k_2 - the maximum number of samples over which the cumulation of defectives (defects) takes place in the second stage of the procedure (also during the normal period).

C_1 - the allowable number of defectives (defects) in the cumulative results from k_1 or fewer samples of n (or n_1). Thus, C_1 is an acceptance number for cumulative results. It is the Cumulative Results Criterion (CRC) that must be met by cumulative sampling results during the first stage of the restart period in order to permit acceptance of a lot.

C_2 - the allowable number of defectives (defects) in the cumulative results from $k_1 + 1$ to k_2 samples of n (or n_2). Thus, C_2 is also an acceptance number for cumulative results and the CRC that must be met by cumulative results during the second stage of the restart period (and also during the normal period) in order to permit acceptance of a lot.

n - the sample size; the number of units (specimens) taken from the lot for inspection.

or n_1 - the sample size in the first stage of the restart period.

n_2 - the sample size in the second stage of the restart period and the normal period.

Other designations helpful in describing the operating procedure are as follows:

d - the number of defectives (defects) in a sample.

d_i - the number of defectives (defects) in the ith sample.

D - the cumulative number of defectives (defects) in a series of samples.

D_i - the cumulative number of defectives (defects) in the ith sample with cumulation performed according to the rules of the plan.

The operating procedure for the general plan is as follows:

Step 1. At the outset (and upon a previous lot rejection), select a random sample of n (or n_1 for the plan with different sample sizes in the two stages) units or specimens from the first lot and from each succeeding lot (up to k_1 lots).

2. Record the number of defectives, d, in each sample and sum the number of defectives, D, in all samples from the first up to and including current sample.

3. Accept the lot associated with each new sample during the cumulation as long as $D_i \leq C_1$; $1 \leq i \leq k_1$. Steps 2 and 3 represent the first stage and the restart period.

4. When k_1 consecutive samples have all resulted in acceptance, continue to sum the defectives, D, in the k_1 samples plus additional samples of n (or n_2 for the plan with different sample sizes in the two stages) up to not more than k_2 samples.

5. Accept the lot associated with each new sample during the cumulation as long as $D_i \leq C_2$; $k_1 < i \leq k_2$. Steps 4 and 5 represent the second stage of the restart period.

6. When the second stage of the restart period has been successfully completed (i.e., k_2 consecutive samples have resulted in acceptance), start cumulation of defectives as a *moving total* over k_2 samples of n (or n_2) by adding the current sample result while dropping from the sum the sample result of the k_2th preceding sample. Continue this procedure as long as $D_i \leq C_2$ and in each instance accept the lot. Step 6 represents the normal period in the second stage.

7. If, for any sample at any stage of the above procedure, D_i is greater than the corresponding C (C_1 or C_2 as the case may be), reject the lot.

8. When a lot is rejected, return to Step 1 and a fresh restart of the cumulation procedure.

An operation schematic for these plans and the above procedure is shown in Figure 3.7.

Specific series of ChSP-C_1,C_2 plans have been studied and presented, particularly in Dodge and Stephens (1964) and Stephens and Dodge (1965a, 1965b, 1966a, 1966b, and 1967). These include sets of C_1,C_2 values of 0,1; 0,2; 1,2; 1,3; 0,4; and 1,4 for the same sample size in the two stages and for different sample sizes in the two stages (Stephens and Dodge (1966b and 1976b). The series affords a wide selection of operating characteristics for choosing plans for specific applications, though no tabulations of plans are yet available (see section 3.3 for work along these lines).

The basic evaluation of these plans is by OC curves, of which many are presented in the reference publications. Comparisons with single and double sampling plans are presented in Stephens and Dodge (1966c) and described in section 3.3. The evaluation of response characteristics is presented in Stephens and Dodge (1967) and is described in section 3.4.

To assist in choosing plans for applications, some formulas for Pa and some OC curves are shown below. The OC curves chosen for presentation are those for large samples and serve (1) to illustrate the general chain sampling effect over that of single sampling, (2) to illustrate the effects of varying the chain sampling parameters, and (3) to provide somewhat of a nomograph that gives suitable approximations for selecting chain sampling plans for sample sizes of n = 50 and larger by using the lower pn scale as illustrated below.

For all ChSP-0,1 with constant n,

$$Pa = \frac{P_0(1 - P_0) + P_1 P_0^{k_1}(1 - P_0^{k_2 - k_1})}{1 - P_0 + P_1 P_0^{k_1}(1 - P_0^{k_2 - k_1 - 1})} \tag{3-8}$$

P_j = the probability of finding j defectives (defects) in a sample of n from product of quality p (j = 0,1). For the special case for which $k_1 = k_2 - 1 = i$, this formula reduces to that of ChSP-1 as given in section 3.1

For ChSP-0,2 and $k_1 = 1$, $k_2 = 2$,

$Pa = P_0/(1 - P_1) + P_0 P_2$, with P_j as above (j = 0,1,2) $\tag{3-9}$

For ChSP-0,2 and $k_1 = 2$, $k_2 = 3$,

$Pa = P_0/(1 - P_0 P_1 - P_0 P_1^2) + P_0^2 P_2$, with P_j as above (j = 0,1,2) $\tag{3-10}$

For ChSP-0,3 and $k_1 = 1$, $k_2 = 2$,

$Pa = (P_0 + P_0 P_2 + P_0 P_3 - P_0 P_1 P_3 - P_0 P_1 P_2 P_3)/(1 - P_1 - P_1 P_2)$, $\tag{3-11}$

with P_j as above (j = 0, 1, 2, 3)

For ChSP-0,4 and $k_1 = 1$, $k_2 = 2$,

27

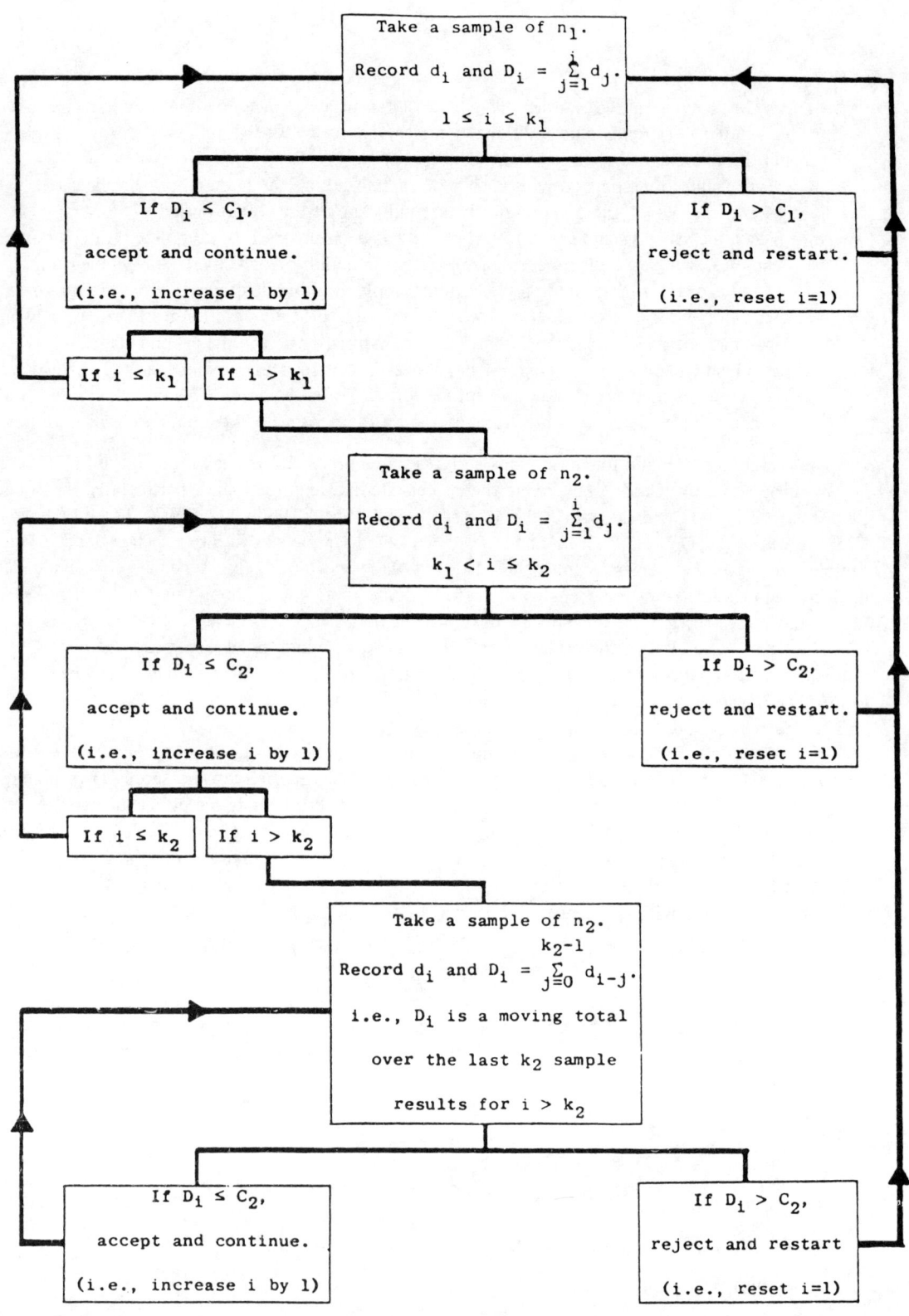

(For the case of equal sample size in the two stages, replace n_1 and n_2 above by n.)

Figure 3.7. Operation Schematic for Generalized Family of Two Stage Chain Sampling Inspection Plans

28

$$Pa = P_0 + P_0P_3 + P_0P_4 + P_0P_1P_4 + P_0P_2P_3 + P_0P_2P_4 + P_0P_1P_3P_4 - P_0P_1P_2P_3P_4)/$$
$$(1-P_1 - P_2 - P_1P_3 + P_1P_2P_3), \text{ with } P_j \text{ as above } (j = 0, 1, 2, 3, 4) \tag{3-12}$$

Results for larger k's, in particular k_2, are too complex for the derivation of algebraic expressions. Consequently, computer analysis is employed using a program to solve the appropriate Markov chain for each plan. (See Stephens and Dodge [1965a and 1974] for details). A set of OC curves from such analysis is shown in Figures 3.8, 3.9, and 3.10 covering a wide range of $k_2 = 2, 3,$ and 5. These OC curves are for $n = 100$, but with the provision of a pn scale, associated with the fraction defective, p, they serve as nomographs for determining the approximate operating characteristics of chain sampling plans having the designated parameters and sample sizes of $n = 50$ and above. The following example illustrates this feature:

For ChSP-0,3 with $k_1, k_2 = 2,3$, curve 5 of the middle curves of Figures 3.9 or 3.10 can be used. Suppose it is desired to determine the OC curve for $n = 60$. List a series of p values of interest. For each value of p, obtain pn by multiplication with n (in this case 60). Enter the pn scale of the appropriate curves with each pn value in the series. Trace up to the OC curve of the desired ChSP and over to the associated Pa scale value, e.g.,

	n = 60	(2,3;0,3)
p	pn	Pa
.005	0.3	.98
.010	0.6	.85
.020	1.2	.42
.030	1.8	.19
:	:	:

For ChSP (n_1,n_2)-0,1 and $k_1 = 1, k_2 = 2$,

$$Pa = (P_{0,n_1} + P_{0,n_1}P_{1,n_2})/(1 + P_{0,n_1} - P_{0,n_2} + P_{0,n_1}P_{1,n_2} - P_{0,n_2}P_{1,n_2}) \tag{3-13}$$

which for $P_{0,n_1} = P_{0,n_2} = P_0$, and $P_{1,n_2} = P_1$

reduces to $Pa = P_0 + P_0P_1$ as expected for ChSP-0,1 or ChSP-1 with $i = 1$.

For ChSP (n_1,n_2)-0,1 and $k_1 = 2, k_2 = 3$,

$$Pa = \frac{(P_{0,n_1} - P_{0,n_1}P_{0,n_2} - P_{0,n_1}P^2_{0,n_2}P_{1,n_2} + P^2_{0,n_1} + P^2_{0,n_1}P_{1,n_2} + P^2_{0,n_1}P_{0,n_2}P_{1,n_2})}{(1 - P_{0,n_2} - P^2_{0,n_2}P_{1,n_2} + P_{0,n_1} - P_{0,n_1}P_{0,n_2} - P_{0,n_1}P_{1,n_2} + P^2_{0,n_1}P_{1,n_2} + P^2_{0,n_1}P_{1,n_2} + P^2_{0,n_1}P_{1,n_2})} \tag{3-14}$$

which also reduces to $Pa = P_0 + P^2_0P_1$, as above, for the constant n result.

For ChSP (n_1,n_2)-0,2 and $k_1 = 1, k_2 = 2$,

$$Pa = \frac{(P_{0,n_1} + P_{0,n_1}P_{2,n_2} - P_{0,n_1}P_{1,n_2}P_{2,n_2})}{(1 + P_{0,n_1} + P_{0,n_1}P_{2,n_2} - P_{0,n_1}P_{1,n_2}P_{2,n_2} - P_{0,n_2} - P_{1,n_2} - P_{0,n_2}P_{2,n_2} + P_{0,n_2}P_{1,n_2}P_{2,n_2})} \tag{3-15}$$

and likewise reduces to the constant n result.

For ChSP (n_1,n_2)-0,3 and $k_1 = 1, k_2 = 2$,

$$Pa = \frac{P_{0,n_1}(1 + P_{2,n_2} + P_{3,n_2} - P_{1,n_2}P_{3,n_2} - P_{1,n_2}P_{2,n_2}P_{3,n_2})}{1 + P_{0,n_1}(1 + P_{2,n_2} + P_{3,n_2} - P_{1,n_2}P_{3,n_2} - P_{1,n_2}P_{2,n_2}P_{3,n_2}) - P_{0,n_2}(1 + P_{2,n_2} + P_{3,n_2} - P_{1,n_2}P_{3,n_2} - P_{1,n_2}P_{2,n_2}P_{3,.2}) - P_{1,n_2} - P_{1,n_2}P_{2,n_2}} \tag{3-16}$$

and also reduces to the constant n result.

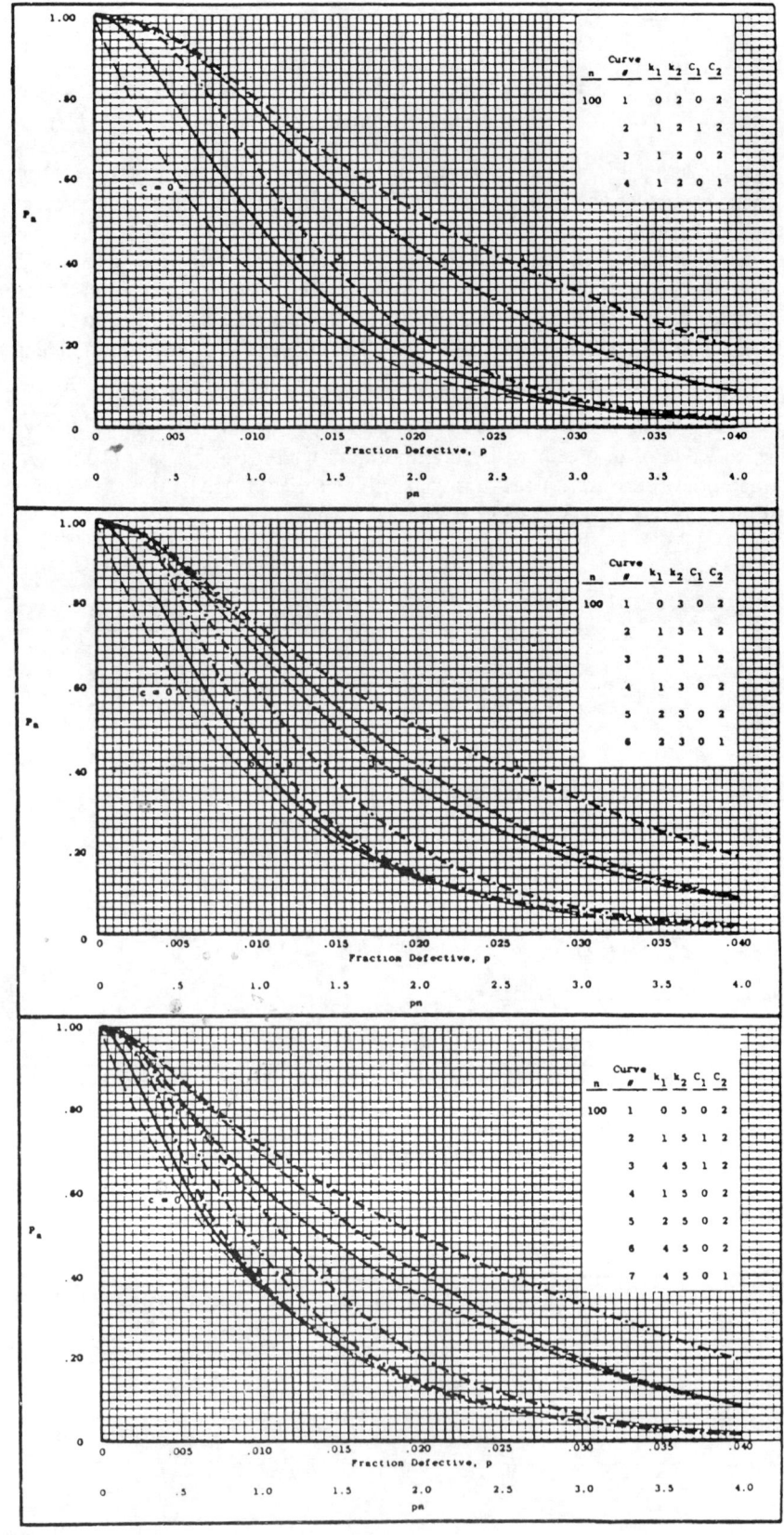

Figure 3.8. OC Curves, ChSP-0,1; 0,2 and 1,2; k_2 = 2,3 and 5; n = 100, with Additional pn Scale

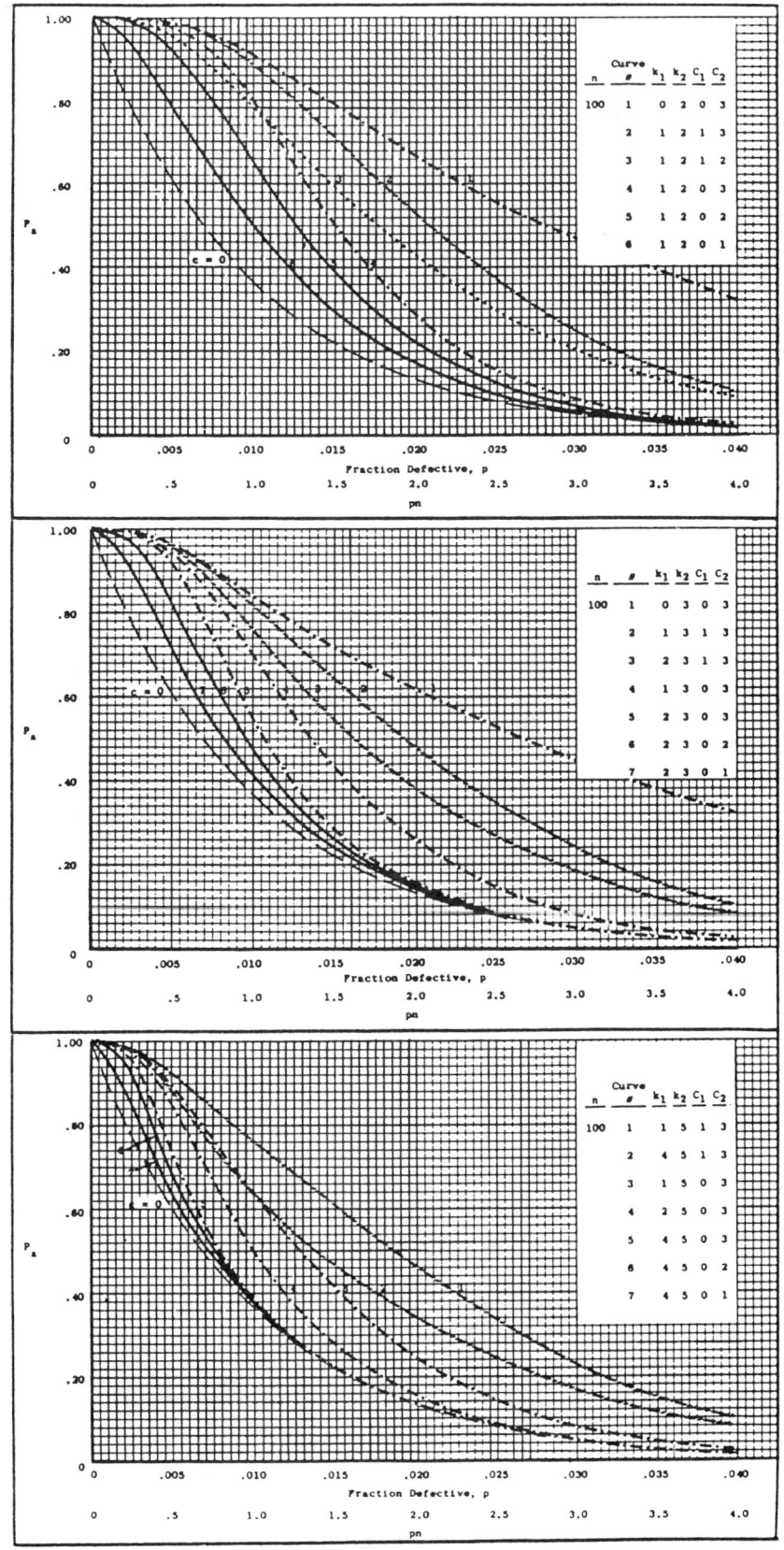

Figure 3.9. OC Curves, ChSP-0,3; 1,3 and with 0,1; 0,2 Comparisons; $k_2 = 2,3$ and 5; $n = 100$, with Additional pn Scale

31

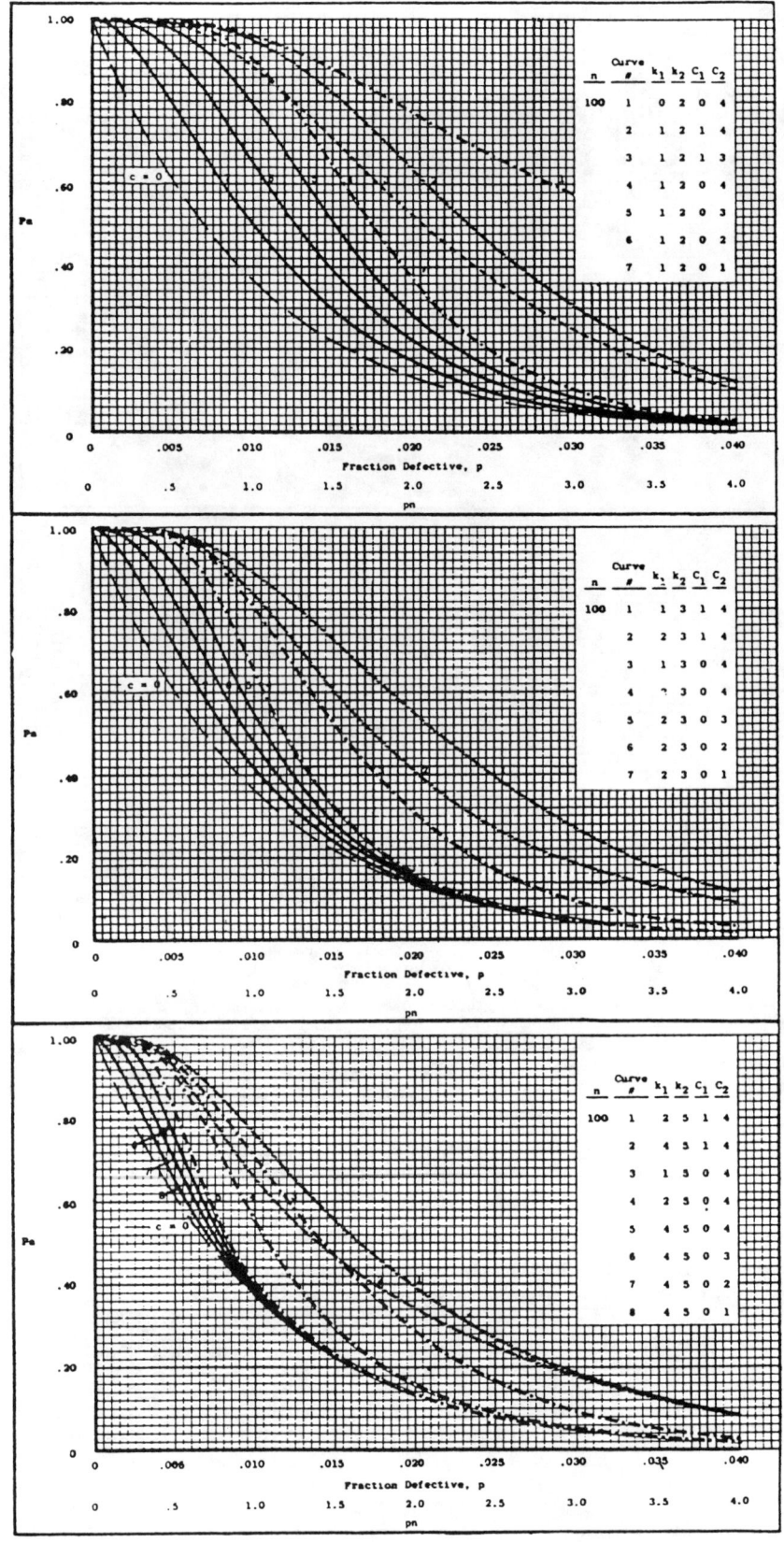

Figure 3.10. OC Curves, ChSP-0,4: 1,4; and with 0,1; 0,2; and 0,3 Comparisons; $k_2 = 2,3$ and 5; n = 100 with Additional pn Scale

32

The plans listed earlier with cumulations over a small number of lots ($k_2 = 2$ and 3) give useful results and are simple to apply. As for ChSP(n_1,n_2)-C_1,C_2, results for larger k_2 yield even more complex algebraic expressions. However, evaluation by computer analysis based on matrix solution of appropriate Markov chains is readily obtained. (See Stephens and Dodge [1966b and 1976b] for details). OC curves from such an analysis are shown in Figures 3.11 and 3.12. They cover a wide range of ChSP(n_1,n_2)-C_1,C_2 plans with C_1, C_2 ranging from 0,1 to 1,3 and $k_2 = 2$, 3, and 5. These OC curves are for $n_1 = 200$, $n_2 = 100$ and also have a pn scale to serve as a nomograph as illustrated above.

Since these plans involve a variable sample size, evaluation of the Average Sample Number (ASN) is useful for additional information about their performance. ASN curves, for the plans with OC curves shown above, are shown in Figures 3.13 and 3.14 respectively. They contain pn and n_2 scales to serve as nomographs.

Further evaluations with respect to the comparison with single and double sampling, including choosing matching ChSP parameters and response operating characteristics, are discussed in sections 3.3 and 3.4 respectively.

While no tables of plans of the generalized family of chain sampling inspection plans are yet available, a large number of plans with evaluations are presented in the references already cited. Additionally, some tabulations of plans are given by Hopkins, Nutt, and Heathcock (1972) in which they report application to acceptance-test firing of missiles. The procedures and evaluations here presented should enable the selection of plans for many applications. Further assistance is provided in section 3.3 by comparing ChSP with single and double sampling plans. The paper by Soundararajan and Govindaraju (1983) should also be consulted for selecting ChSP procedures based on three sets of criteria.

3.3 COMPARISONS WITH SINGLE AND DOUBLE SAMPLING PLANS - ChSP CHOICES

Comparisons of chain sampling plans with single and double sampling plans are presented by Stephens and Dodge (1966c and 1976a). The approach is by means of the operating ratio, OR, as illustrated near the end of section 2.2.1 for skip-lot sampling. A table of OR values for ChSP, SS, and DS plans is shown in Figure 3.15. OR values for the ChSP plans are based on the operating characteristics for n = 50. Those for the SS and DS plans are based on the operating characteristics for $n_2 - 2n_1$ and are from Tables 7.1 and 8.2 of Duncan (1986).

Based on the potential matching illustrated by the results of Figure 3.15 some specific matches are shown in Figures 3.16 and 3.17.

These results illustrate the greater efficiency of chain sampling plans from an operating characteristic viewpoint. The results presented are for the smallest of cumulation parameters, viz $k_1,k_2 = 1,2$. An even greater range of potential matching with single and especially double sampling plans is possible for ChSP plans with other k_1,k_2 combinations. Figure 3.18 presents a table of approximate OR values for a limited but useful set of ChSP parameters. Approximate values of $np_{.95}$ are also presented for use in deriving chain sampling plans having desired operating characteristics.

Use of this table to derive a ChSP plan having desired properties is illustrated by the following example.

A chain sampling plan is to be devised. Desired operating characteristics include a satisfactory producer's quality level of 1.5% defectives having a 95% probability of acceptance (AQL = 1.5%) and a limiting quality level of 8% defective having a 10% probability of acceptance (LQL = 8%). These values can be expressed as, $p_{.95} = 0.015$, $p_{.10} = 0.08$ respectively. Hence the desired OR = $p_{.10}/p_{.95}$ = 0.08/0.015 = 5.33. This corresponds approximately to the OR of the plan $(k_1,k_2;C_1,C_2) = (2,3;0,3)$ in Figure 3.18. For this plan, $np_{.95}$ is shown as 0.43; hence the sample size is obtained by solution of the equation, $np_{.95} = 0.43$ or $n = 0.43/p_{.95} = 0.43/0.015$ = 28.66 or 29. The ChSP plan having approximately the desired properties is $(n;k_1,k_2;C_1,C_2) = (29;2,3;0,3)$.

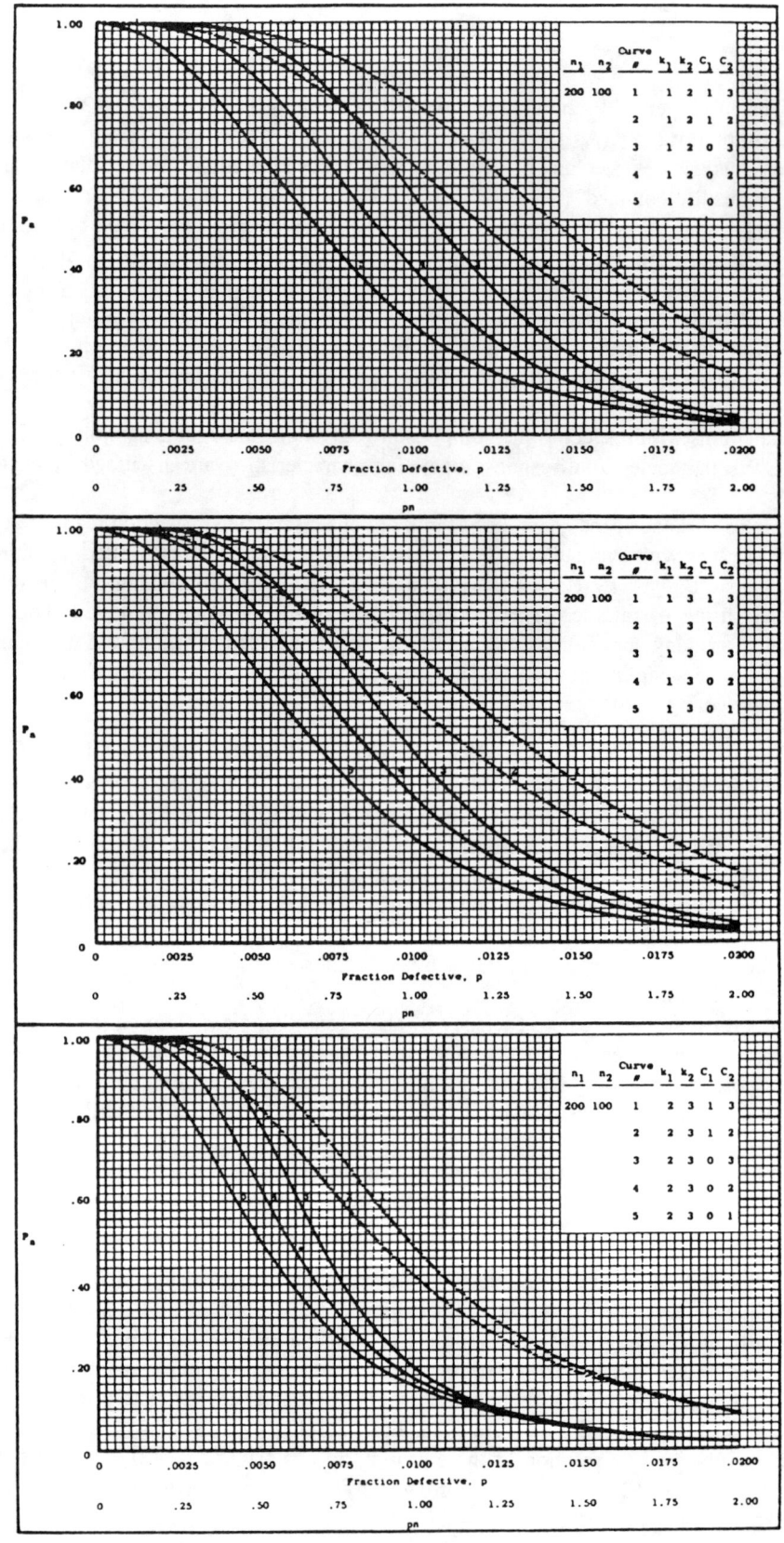

Figure 3.11. OC Curves, ChSP (200, 100) - 0,1; 0,2; 1,2; 0,3; 1,3; k_2 = 2 and 3, with Additional pn Scale

34

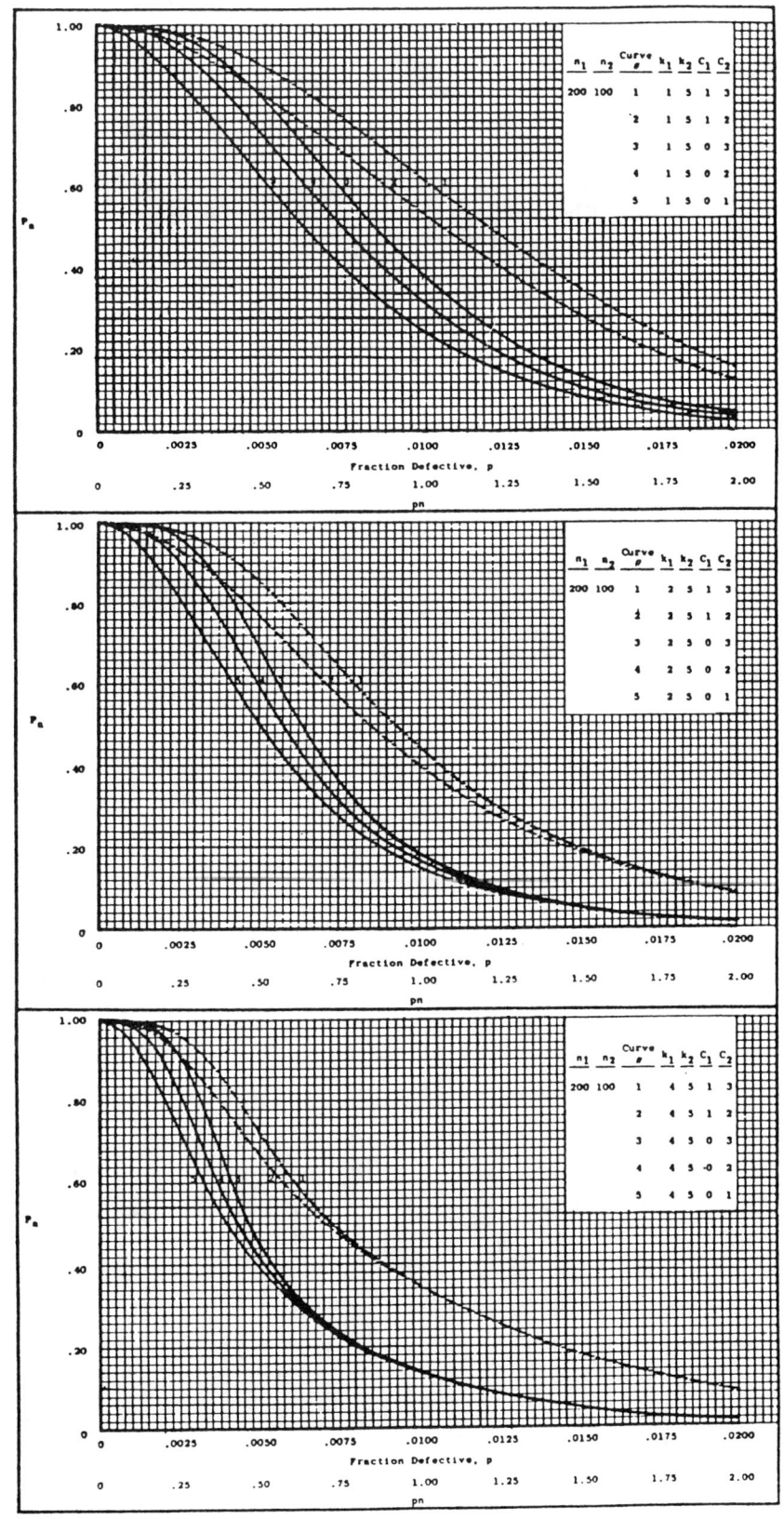

Figure 3.12. OC Curves, ChSP (200, 100) - 0,1; 0,2; 1,2; 0,3; 1,3; $k_2 = 5$, with Additional pn Scale

35

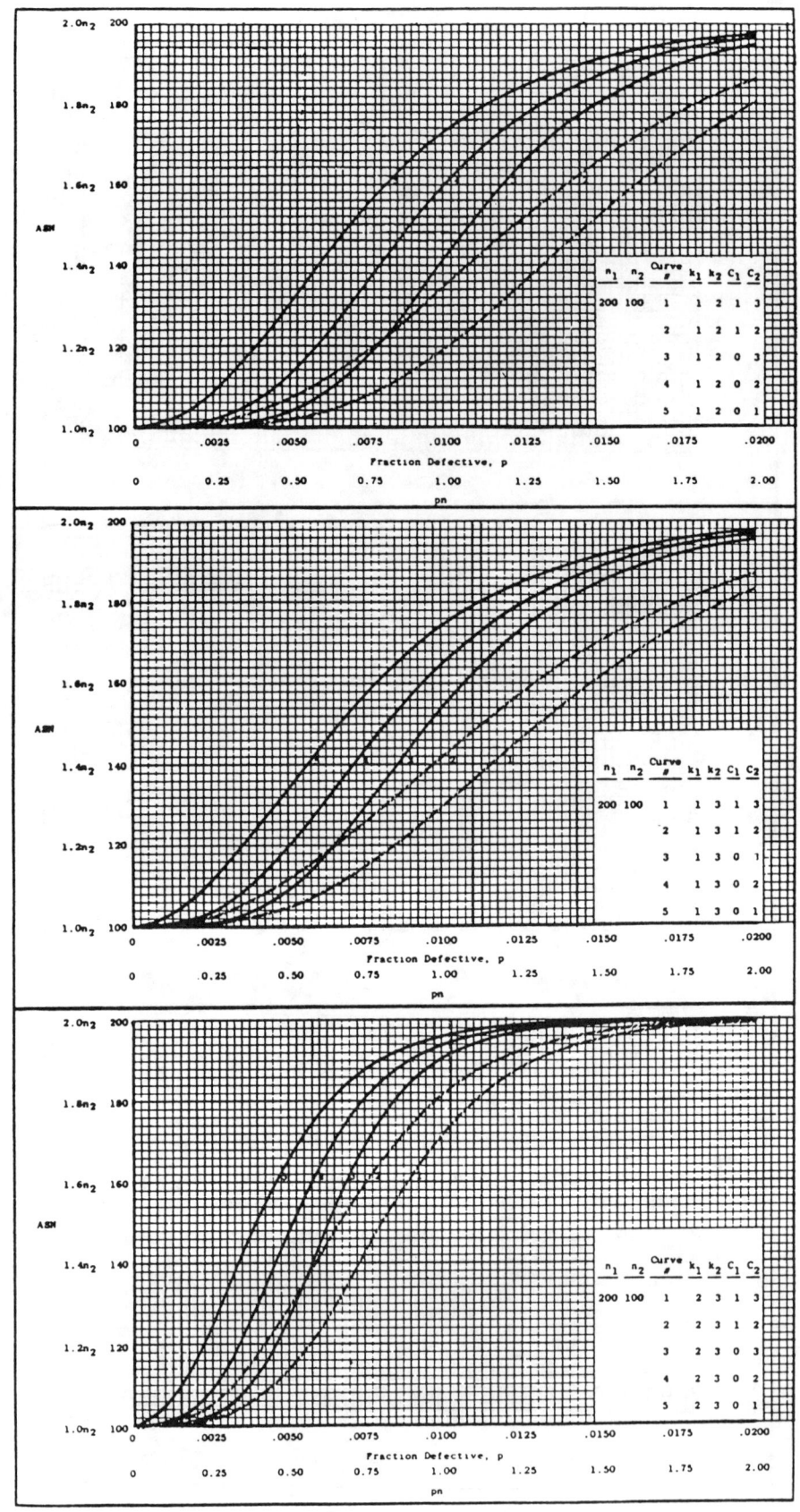

Figure 3.13. ASN Curves, ChSP (200, 100) - 0,1; 0,2; 1,2; 0,3; 1,3; k_2 = 2 and 3, with Additional pn and n_2 Scales

36

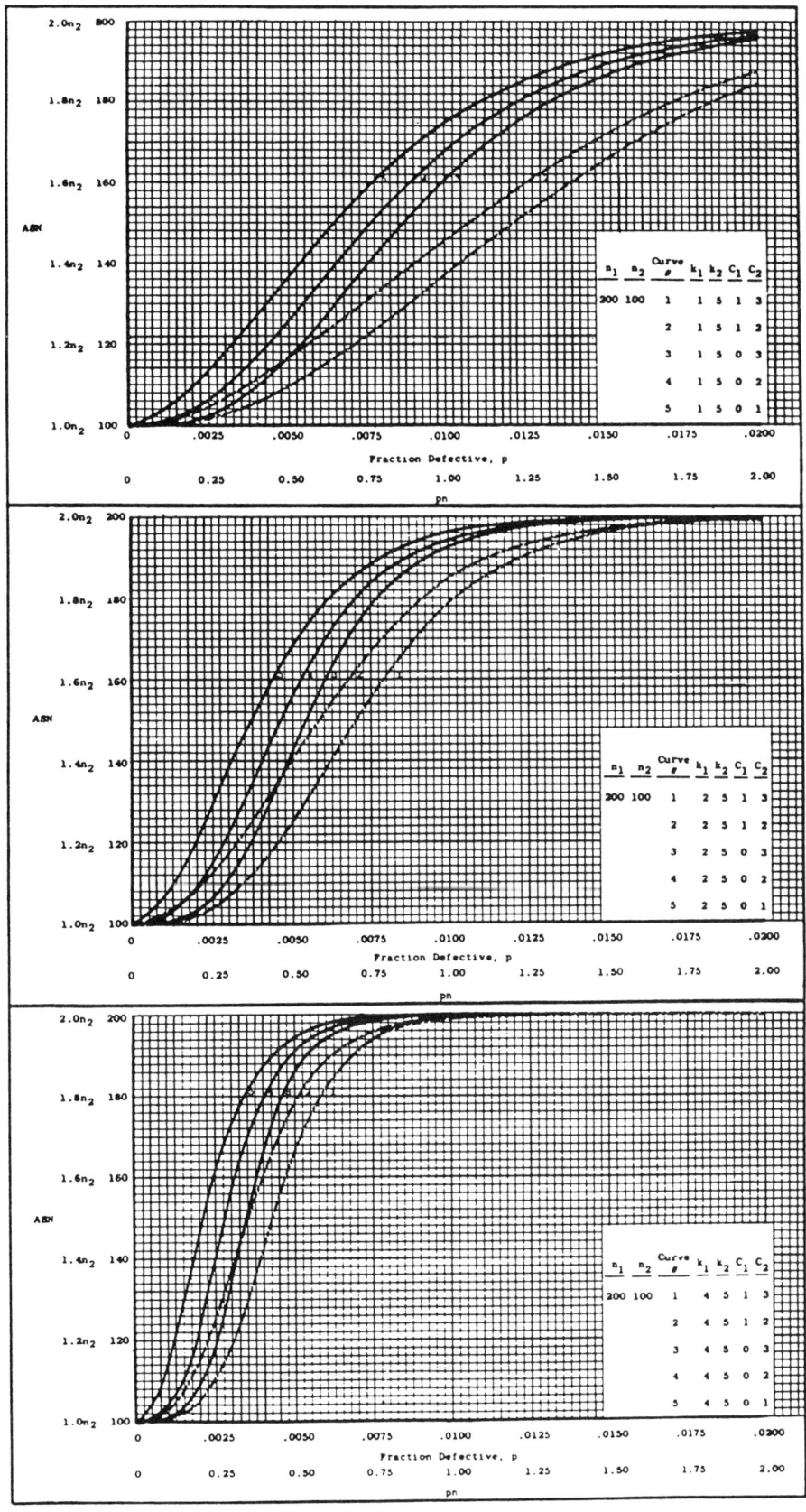

Figure 3.14. ASN Curves, ChSP (200, 100) - 0,1; 0,2; 1,2; 0,3; 1,3; $k_2 = 5$, with Additional pn and n_2 Scales

37

ChSP Plans			SS Plans		DS Plans	
k_1,k_2	C_1,C_2	OR	c	OR	C_1,C_2	OR
1,2	0,1	12.4	1	10.96	0,1	14.50
	1,2	8.4				
	0,2	6.6	2	6.50	1,3	6.48
	1,3	5,4				
	0,3	4.5	3	4.89	0,4	4.31
	1,4	4.1				
	0,4	3.5	5	3.55	1,6	3.60

Figure 3.15. Comparison of OR Values for Some ChSP, SS, and DS Plans

For matching of this plan with single and/or double sampling, the OR tables of Duncan (1986) reveal the following:

Single Sampling:	c	OR	$np_{.95}$	$n = np_{.95}/p_{.95}$
(1)	2	6.50	0.818	55
(2)	3	4.89	1.366	91

Since the desired OR = 5.33, the first plan with n = 55, c = 2 will be less discriminating (larger OR) while the second plan with n = 91, c = 3 will be more discriminating (smaller OR) than the ChSP plan.

Double Sampling:

(1) For $n_1 = n_2$, $c_1 = 1$, $c_2 = 3$, OR = 5.39, $n_1 p_{.95} = 0.76$;
hence, $n_1 = n_2 = 0.76/0.015 \approx 51$.

	ChSP	SS	DS				ChSP	SS	DS	
	n=50 1,2;0,1	n=85 c=1	n_1=50, n_2=50 c_1=0 c_2=1		n_1=40, n_2=80 c_1=0 c_2=1		n=50 1,2;0,2	n=105 c=2	n_1=72 n_2=144 c_1=1 c_2=3	
p	Pa	Pa	Pa	ASN	Pa	ASN	Pa	Pa	Pa	ASN
.002	.987	.987	.987	54.5	.986	45.9	.999	.999	.999	73.3
.005	.931	.932	.931	59.8	.928	53.2	.986	.984	.989	79.2
.008	.850	.852	.850	63.5	.848	58.7	.952	.947	.955	88.0
.010	.790	.791	.790	65.3	.790	61.6	.917	.911	.918	94.5
.015	.638	.635	.638	67.9	.646	66.6	.794	.791	.787	111.0
.020	.499	.491	.499	68.6	.518	69.1	.647	.649	.637	124.8
.025	.384	.370	.384	68.1	.412	69.8	.506	.510	.497	134.5
.030	.292	.273	.292	66.9	.328	69.3	.385	.387	.381	139.6
.040	.165	.141	.165	63.5	.208	66.0	.214	.204	.217	138.7
.050	.093	.070	.093	60.1	.133	61.6	.117	.099	.120	128.6
.060	.052	.033	.052	57.2	.086	57.2	.063	.045	.065	115.4

Figure 3.16. ChSP-0,1 and ChSP-0,2 Plans Matched With SS and DS Plans

38

	ChSP	SS	DS		ChSP	SS		DS	
	n=50 $1,2;0,3$	n=120 c=3	$n_1=54,$ $c_1=0$	$n_2=108$ $c_2=4$	n=50, $1,2;0,4$	$n_2=135$ c=4	n=155 c=5	$n_1=69$ $c_1=1$	$n_2=138$ $c_2=6$
p	Pa	Pa	Pa	ASN	Pa	Pa	Pa	Pa	ASN
.002	.999	.999	.999	65.1	.999	.999	.999	.999	70.2
.005	.998	.997	.999	79.6	.999	.999	.999	.999	75.5
.008	.989	.984	.991	92.0	.998	.995	.998	.999	83.6
.010	.977	.967	.979	99.2	.995	.988	.995	.996	89.9
.015	.911	.893	.912	114.1	.971	.946	.970	.971	107.2
.020	.794	.780	.797	125.2	.906	.865	.908	.903	124.5
.025	.648	.647	.653	133.3	.792	.750	.807	.789	140.2
.030	.502	.513	.506	138.7	.645	.619	.678	.648	153.5
.040	.277	.289	.268	143.1	.365	.369	.410	.376	172.2
.050	.147	.144	.128	141.0	.188	.190	.208	.192	180.6
.060	.078	.066	.059	134.2	.096	.087	.092	.094	180.1

Figure 3.17. ChSP-0,3 and ChSP-0,4 Plans Matched With SS and DS Plans

(2) for $n_2 = 2n_1$, $c_1 = 0$, $c_2 = 3$, OR = 5.39, $n_1p_{.95} = 0.49$; hence, $n_1 = 0.49/0.015 \approx 33$ and $n_2 = 66$. The chain sampling plan involves a considerably smaller sample size.

3.4 EVALUATION OF RESPONSE CHARACTERISTICS

As indicated in section 3.1, the OC curves of ChSP procedures have different meanings than those of ordinary lot-by-lot sampling plans. Analyses by means of OC curves, in general, use a theoretical model in which the process quality is stable - that is, one having constant p. However, in practice, process quality levels are subject to changes such as trends, sudden shifts, erratic variations, etc. For ordinary lot-by-lot sampling procedures the entire sample results for making an acceptance-rejection decision come from the lot under consideration. Hence, the probability of acceptance of a given lot is determined by the quality of the lot, independent of the quality of neighboring (e.g., preceding) lots. Chain sampling, on the other hand, uses cumulative sample results from preceding lots in sentencing a given lot. Hence, the probability of acceptance of a current lot is dependent on the quality of the current lot *as well as* the quality(ies) of the preceding lots whose sample results are used in the cumulative results. Therefore, the comparisons made in section 3.3 are valid under the assumption of a constant p process quality model. The conditions to be satisfied for applications of chain sampling, as outlined near the end of section 3.1, are intended to enhance the validity of that assumption.

k_1,k_2 \ C_1,C_2	0,1		0,2		0,3		0,4	
	OR	$np_{.95}$	OR	$np_{.95}$	OR	$np_{.95}$	OR	$np_{.95}$
1,2	11.7	.21	6.5	.41	4.6	.62	3.5	.85
1,3	14.5	.17	8.3	.32	5.8	.48	4.6	.64
2,3	14.4	.16	8.0	.29	5.5	.43	4.2	.56

Figure 3.18. Operating Ratios and $np_{.95}$ Values for Some Useful Chain Sampling Plans

However, even under the best operating conditions, process quality levels may change on occasion and it is well to investigate how chain sampling procedures behave, at least under the model of an abrupt change in level—an event to be detected and corrected as early as possible. Such an evaluation is carried out by Stephens and Dodge (1967). Only the principle conclusions are summarized here. This report should be consulted for additional details.

Evaluation is by means of the Average Run Length, ARL, to a rejection. Three measures of ARL are defined and used in the evaluation, with the following two of particular significance,

ARL-1   is the average number of samples (hence lots) to the first occurrence of a rejection (R) after the process quality shifts abruptly, at a random point, from a level of p_0 to a level of p_1 ($p_1 > p_0$), with the count beginning on the first sample (lot) after the shift.

ARL-2   is the average number of samples to the first occurrence of a rejection (R) following a rejection, with the process quality at a level of p.

Derivation of a general result for these ARL's is developed using the Markov chain structure.

ARL-2 is also the average run length of the first rejection for a non-cumulative type sampling plan and equals $1/(1-Pa)$. Thus, for a given cumulative result plan, the comparison, ARL-1 versus ARL-2, is a comparison of the behavior of that particular plan with the behavior of a non-cumulative plan having the *same* operating characteristic (OC) curve.

An examination of these measures, ARL-1 and ARL-2, and their differences for several sets of parameters results in the following.

(1)   For a fixed k_2 for different k_1's from 0 to k_2-1, the values of ARL-1 for a given p_0 do not differ much no matter what the value of p_1 (the differences are too small to show on a chart using scales that are suitable for plotting ARL-1 versus p_1).

(2)   For increasing k_2, the values of ARL-1 are reduced, but at a decreasing rate as k_2 gets larger. This can be seen in the ARL-1 curves shown in Figure 3.19. The curves plotted are those of the designated k_2 with $k_1 = 0$, since these are sufficient to give a general picture for each k_2 for reasons noted in (1) above. Separate sets of curves are given for the ChSP-0,1 and ChSP-0,2 subsets. In each case the dashed curve represents the ARL of the single sample plan: n = 50, c = 0, which is shown for comparison. A value of $p_0 = p_{.98}$ is chosen as a reference quality level, resulting in an ARL = $1/.02 = 50$.

(3)   For increasing k_2, and also for increasing k_1 within k_2, the values of ARL-2 are reduced considerably while those for ARL-1 are reduced only slightly - making the difference, (ARL-1)-(ARL-2), increase considerably. This reflects the tightening of the OC curves with increasing k_2 and for increasing k_1 within k_2, since ARL-2 = $1/(1-Pa)$.

(4)   ARL-1 is increased considerably by increases in C_2. This can be seen between the two graphs of Figure 3.19.

These results emphasize that while we can tighten up the OC curves by increasing k_1, we can achieve in response sensitivity practically nothing more than that of the $k_1 = 0$ plan for a given value of k_2, if concern is in detecting the first rejection after a shift in quality level from low values of p (say $p_{.98}$). That is, ARL-1 is dominated by k_2. But it does not mean that we ought to use the $k_1 = 0$ plan that has a poor OC curve. For most practical applications, we also want a plan that will continue to tell us repeatedly (e.g., after the first rejection) that we are at an undesirable level of quality. For this, we will want a plan that has an OC curve as tight and as discriminating as possible, knowing that its ability to detect shifts as measured by the first rejection after the shift (from small p_0 to large p_1) is only as good as its $k_1 = 0$, given k_2, plan (as shown in Figure 3.19).

Dominance of ARL-1 by second stage criteria (k_2 and C_2) leads to still additional important points, viz.

(1)   Little is gained, *with respect to response sensitivity,* by using $n_1 = kn_2$ and in particular $n_1 = 2n_2$ as considered in the examples of ChSP(n_1,n_2)-C_1,C_2 plans of section 3.2.2 and in

40

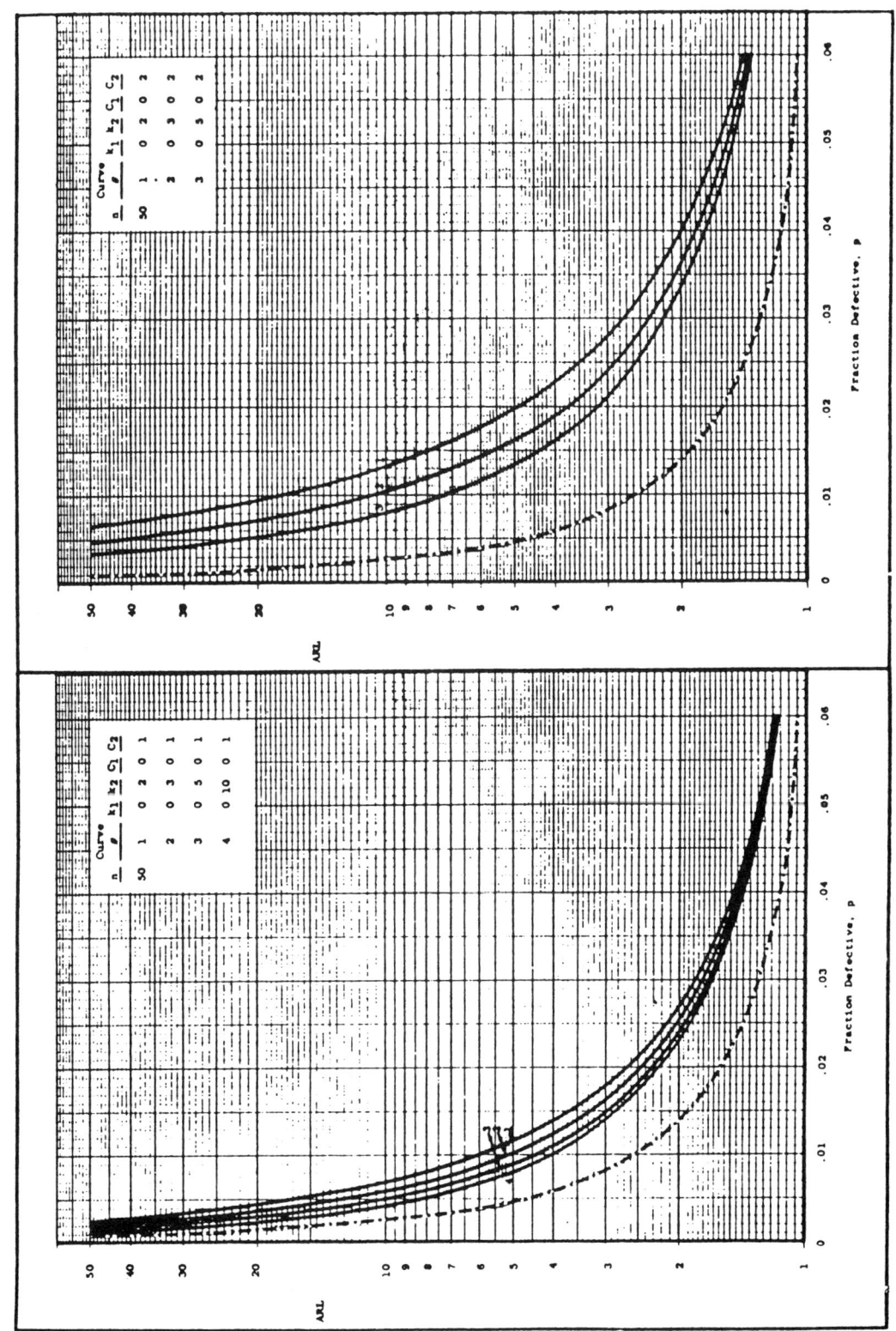

Figure 3.19. ARL-1 Curves of Some ChSP-0,1 and ChSP-0,2 Plans

Stephens and Dodge (1966b and 1976b). However, the OC curve discussion above is still applicable.

(2) Since the difference, (ARL-1) - (ARL-2), is generally smallest for the $k_1 = 0$ plans, *an approximation for ARL-1*, for a given plan, is provided by ARL-2 (and hence the operating characteristics, since ARL-2 = 1/(1-Pa) of the $k_1 = 0$, given k_2, plan.

In further summary, most ChSP plans are seen to have a larger ARL-1 than equivalent (OC curve-wise) non-cumulative plans (ARL-1 = ARL-2 for non-cumulative plans). This difference or increase in ARL measures the extra lag for the ChSP plans and is one measure of the price to be paid for the greater efficiency - less inspection for a given OC curve. The magnitude of this extra lag is sharply controlled by the second stage parameters, k_2 and C_2, but especially C_2. A larger C_2, other things equal, means a greater lag, although achieving improved discrimination in the OC curve. Because of this, a larger sample in the first stage, which also improves the OC curves, has little influence on ARL-1. And a fairly good approximation of ARL-1 (which requires extensive numerical analysis to determine) is provided by ARL-2 of the $(0, k_2; C_1, C_2)$ plan.

It should be noted that because of the restart feature of ChSP plans, the extra lag discussed above occurs only once, only until the first rejection. After the first rejection, for the model being considered, the spacing between rejections is controlled wholly by the OC curve, the same as for any non-cumulative plan. When ChSP plans are applied in the shop, the actual lag is greatest percentage-wise for minor process shifts. The actual lag associated with substantial shifts is relatively small. Thus, it is believed that the gain in inspection efficiency (reduction in ASN) for two-stage chain sampling plans more than offsets the extra lag in detecting changes in quality. This is particularly true if the conditions of application (section 3.1) are satisfied - reducing the necessity to detect changes in quality.

APPENDIX A

VALUES OF Y FOR SkSP-2 SKIP-LOT PLANS

TO DETERMINE P_L^* IN: $P_L = Y/n$

$n/N = 0$ (Large N)

c	f	i			
		4	6	8	10
1	2/3	0.8682	0.8479	0.8421	0.8405
	1/2	0.8954	0.8564	0.8443	0.8411
	1/3	0.9443	0.8784	0.8493	0.8423
	1/4	0.9861	0.8939	0.8549	0.8436
	1/5	1.0219	0.9125	0.8613	0.8450
2	2/3	1.4281	1.3935	1.3794	1.3741
	1/2	1.4785	1.4163	1.3884	1.3773
	1/3	1.5619	1.4604	1.4081	1.3844
	1/4	1.6284	1.5000	1.4291	1.3927
	1/5	1.6835	1.5349	1.4501	1.4021
3	2/3	2.0294	1.9835	1.9610	1.9505
	1/2	2.1013	2.0229	1.9806	1.9593
	1/3	2.2177	2.0927	2.0205	1.9789
	1/4	2.3067	2.1511	2.0582	2.0004
	1/5	2.3971	2.2006	2.0925	2.0223
4	2/3	2.6604	2.6054	2.5754	2.5594
	1/2	2.7547	2.6615	2.6076	2.5764
	1/3	2.8998	2.7561	2.6683	2.6124
	1/4	3.0097	2.8320	2.7217	2.6482
	1/5	3.0980	2.8948	2.7681	2.6817
5	2/3	3.3140	3.2516	3.2151	3.1939
	1/2	3.4286	3.3242	3.2605	3.2207
	1/3	3.6018	3.4423	3.3417	3.2742
	1/4	3.7312	3.5346	3.4098	3.3236
	1/5	3.8344	3.6100	3.4674	3.3677

$*P_L = AOQL_1$ (See Section 2.2.1)

n/N = 0 (Large N)

c	f	i			
		4	6	8	10
6	2/3	3.9857	3.9171	3.8751	3.8491
	1/2	4.1197	4.0058	3.9338	3.8866
	1/3	4.3195	4.1463	4.0347	3.9575
	1/4	4.4673	4.2543	4.1168	4.0198
	1/5	4.5846	4.3416	4.1850	4.0738
7	2/3	4.6726	4.5986	4.5518	4.5215
	1/2	4.8250	4.7029	4.6237	4.5700
	1/3	5.0502	4.8650	4.7437	4.6580
	1/4	5.2156	4.9878	4.8390	4.7326
	1/5	5.3462	5.0865	4.9174	4.7959
8	2/3	5.3722	5.2937	5.2425	5.2083
	1/2	5.5424	5.4130	5.3275	5.2680
	1/3	5.7919	5.5958	5.4658	5.3727
	1/4	5.9740	5.7330	5.5739	5.4590
	1/5	6.1174	5.8425	5.6620	5.5313
9	2/3	6.0829	6.0003	5.9454	5.9077
	1/2	6.2702	6.1343	6.0431	5.9785
	1/3	6.5430	6.3372	6.1993	6.0993
	1/4	6.7413	6.4881	6.3197	6.1970
	1/5	6.8969	6.6080	6.4171	6.2779
10	2/3	6.8033	6.7172	6.6588	6.6180
	1/2	7.0070	6.8654	6.7690	6.7000
	1/3	7.3024	7.0876	6.9425	6.8363
	1/4	7.5162	7.2518	7.0748	6.9450
	1/5	7.6836	7.3819	7.1812	7.0341

n/N = 0.10

c	f	\multicolumn{4}{c}{i}			
		4	6	8	10
1	2/3	0.7878	0.7649	0.7583	0.7566
	1/2	0.8910	0.7749	0.7609	0.7572
	1/3	0.8754	0.7975	0.7668	0.7586
	1/4	0.9226	0.8213	0.7742	0.7602
	1/5	0.9626	0.8441	0.7832	0.7619
2	2/3	1.3007	1.2605	1.2438	1.2375
	1/2	1.3603	1.2888	1.2550	1.2414
	1/3	1.4577	1.3441	1.2813	1.2506
	1/4	1.5340	1.3929	1.3102	1.2626
	1/5	1.5962	1.4351	1.3385	1.2776
3	2/3	1.8533	1.7988	1.7711	1.7581
	1/2	1.9414	1.8488	1.7969	1.7696
	1/3	2.0783	1.9375	1.8515	1.7978
	1/4	2.1817	2.0098	1.9025	1.8305
	1/5	2.2643	2.0697	1.9475	1.8632
4	2/3	2.4343	2.3677	2.3300	2.3095
	1/2	2.5505	2.4407	2.3737	2.3330
	1/3	2.7253	2.5620	2.4576	2.3866
	1/4	2.8543	2.6567	2.5294	2.4398
	1/5	2.9562	2.7333	2.5899	2.4877
5	2/3	3.0369	2.9601	2.9135	2.8855
	1/2	3.1807	3.0562	2.9766	2.9242
	1/3	3.3923	3.2095	3.0895	3.0045
	1/4	3.5458	3.3257	3.1812	3.0769
	1/5	3.6659	3.4183	3.2564	3.1391

n/N = .10

c	f	i			
		4	6	8	10
6	2/3	3.6569	3.5711	3.5166	3.4817
	1/2	3.8278	3.6905	3.5998	3.5372
	1/3	4.0750	3.8750	3.7413	3.6443
	1/4	4.2522	4.0120	3.8522	3.7352
	1/5	4.3898	4.1200	3.9416	3.8108
7	2/3	4.2914	4.1976	4.1360	4.0946
	1/2	4.4890	4.3401	4.2396	4.1679
	1/3	4.7708	4.5552	4.4092	4.3015
	1/4	4.9708	4.7125	4.5388	4.4102
	1/5	5.1253	4.8354	4.6420	4.4989
8	2/3	4.9381	4.8371	4.7690	4.7218
	1/2	5.1619	5.0026	4.8932	4.8134
	1/3	5.4777	5.2478	5.0905	4.9731
	1/4	5.6998	5.4248	5.2383	5.0992
	1/5	5.8706	5.5622	5.3549	5.2006
9	2/3	5.5954	5.4879	5.4139	5.3611
	1/2	5.8451	5.6762	5.5586	5.4714
	1/3	6.1940	5.9510	5.7832	5.6570
	1/4	6.4378	6.1473	5.9490	5.8001
	1/5	6.6244	6.2989	6.0786	5.9139
10	2/3	6.2621	6.1485	6.0689	6.0112
	1/2	6.5372	6.3594	6.2342	6.1402
	1/3	6.9187	6.6634	6.4858	6.3513
	1/4	7.1836	6.8787	6.6692	6.5111
	1/5	7.3856	7.0440	6.8116	6.6371

n/N = .2

c	f	i			
		4	6	8	10
1	2/3	0.7078	0.6820	0.6746	0.6727
	1/2	0.7438	0.6939	0.6776	0.6734
	1/3	0.8081	0.7222	0.6849	0.6750
	1/4	0.8612	0.7520	0.6953	0.6768
	1/5	0.9056	0.7797	0.7093	0.6790
2	2/3	1.1741	1.1281	1.1084	1.1009
	1/2	1.2439	1.1632	1.1224	1.1056
	1/3	1.3564	1.2322	1.1588	1.1183
	1/4	1.4427	1.2914	1.1987	1.1384
	1/5	1.5121	1.3412	1.2355	1.1635
3	2/3	1.6785	1.6152	1.5819	1.5660
	1/2	1.7832	1.6784	1.6162	1.5812
	1/3	1.9429	1.7888	1.6908	1.6242
	1/4	2.0608	1.8761	1.7577	1.6739
	1/5	2.1538	1.9467	1.8142	1.7196
4	2/3	2.2099	2.1321	2.0862	2.0603
	1/2	2.3498	2.2250	2.1454	2.0935
	1/3	2.5558	2.3765	2.2585	2.1739
	1/4	2.7043	2.4910	2.3511	2.2494
	1/5	2.8196	2.5815	2.4263	2.3133
5	2/3	2.7621	2.6715	2.6145	2.5788
	1/2	2.9372	2.7951	2.7007	2.6348
	1/3	3.1887	2.9873	2.8520	2.7526
	1/4	3.3667	3.1283	2.9695	2.8524
	1/5	3.5037	3.2382	3.0628	2.9338

(continued)

n/N = .2

c	f	i			
		4	6	8	10
6	2/3	3.3309	3.2290	3.1620	3.1172
	1/2	3.5412	3.3837	3.2763	3.1984
	1/3	3.8375	3.6162	3.4655	3.3531
	1/4	4.0443	3.7832	3.6074	3.4765
	1/5	4.2022	3.9120	3.7184	3.5748
7	2/3	3.9135	3.8014	3.7523	3.6722
	1/2	4.1589	3.9873	3.8683	3.7799
	1/3	4.4994	4.2599	4.0952	3.9711
	1/4	4.7344	4.4526	4.2612	4.1176
	1/5	4.9129	4.5997	4.3894	4.2325
8	2/3	4.5078	4.3864	4.3020	4.2412
	1/2	4.7883	4.6038	4.4741	4.3762
	1/3	5.1724	4.9161	4.7384	4.6035
	1/4	5.4350	5.1340	4.9281	4.7729
	1/5	5.6329	5.2992	5.0733	4.9040
9	2/3	5.1123	4.9822	4.8902	4.8223
	1/2	5.4276	5.2312	5.0915	4.9850
	1/3	5.8549	5.5831	5.3932	5.2483
	1/4	6.1446	5.8257	5.6064	5.4402
	1/5	6.3620	6.0087	5.7683	5.5874
10	2/3	5.7256	5.5874	5.4883	5.4139
	1/2	6.0758	5.8681	5.7192	5.6046
	1/3	6.5459	6.2593	6.0580	5.9037
	1/4	6.8622	6.5265	6.2944	6.1180
	1/5	7.0986	6.7269	6.4728	6.2810

(continued)

n/N = .5

c	f	i 4	6	8	10
1	2/3	0.4717	0.4350	0.4236	0.4209
	1/2	0.5277	0.4603	0.4289	0.4219
	1/3	0.6209	0.5209	0.4607	0.4247
	1/4	0.6921	0.5277	0.5000	0.4506
	1/5	0.7490	0.6143	0.5336	0.4784
2	2/3	0.8032	0.7391	0.7051	0.6918
	1/2	0.9117	0.8112	0.7476	0.7053
	1/3	1.0742	0.9347	0.8462	0.7831
	1/4	1.1907	1.0265	0.9236	0.8507
	1/5	1.2808	1.0980	0.9845	0.9045
3	2/3	1.1673	1.0817	1.0274	0.9943
	1/2	1.3318	1.2050	1.1210	1.0597
	1/3	1.5651	1.3930	1.2813	1.2003
	1/4	1.7266	1.5255	1.3969	1.3044
	1/5	1.8491	1.6263	1.4851	1.3842
4	2/3	1.5546	1.4509	1.3810	1.3312
	1/2	1.7773	1.6277	1.5267	1.4517
	1/3	2.0823	1.8817	1.7497	1.6532
	1/4	2.2885	2.0552	1.9040	1.7944
	1/5	2.4427	2.1849	2.0195	1.9003
5	2/3	1.9596	1.8400	1.7570	1.6951
	1/2	2.2421	2.0722	1.9562	1.8692
	1/3	2.6195	2.3933	2.2431	2.1325
	1/4	2.8700	2.6078	2.4363	2.3112
	1/5	3.0553	2.7663	2.5791	2.4433

c	f	i			
		4	6	8	10
6	2/3	2.3788	2.2447	2.1503	2.0785
	1/2	2.7223	2.5339	2.4043	2.3065
	1/3	3.1725	2.9230	2.7562	2.6328
	1/4	3.4671	3.1786	2.9886	2.8492
	1/5	3.6831	3.3656	3.1585	3.0075
7	2/3	2.8098	2.6625	2.5577	2.4771
	1/2	3.2154	3.0098	2.8676	2.7598
	1/3	3.7387	3.4676	3.2854	3.1501
	1/4	4.0771	3.7642	3.5570	3.4045
	1/5	4.3234	3.9796	3.7540	3.5890
8	2/3	3.2508	3.0912	2.9768	2.8882
	1/2	3.7192	3.4978	3.3437	3.2267
	1/3	4.3159	4.0248	3.8282	3.6817
	1/4	4.6979	4.3642	4.1392	3.9743
	1/5	4.9743	4.6060	4.3632	4.1849
9	2/3	3.7004	3.5292	3.4059	3.3099
	1/2	4.2324	3.9960	3.8309	3.7052
	1/3	4.9028	4.5927	4.3826	4.2256
	1/4	5.3282	4.9714	4.7330	4.5566
	1/5	5.6344	5.2429	4.9839	4.7932
10	2/3	4.1575	3.9755	3.8437	3.7409
	1/2	4.7538	4.5033	4.3277	4.1937
	1/3	5.4980	5.1701	4.9472	4.7802
	1/4	5.9667	5.5898	5.3371	5.1496
	1/5	6.3024	5.8891	5.6147	5.4122

APPENDIX B

NOTE ON CSP-1 AND SkSP-1

EQUATIONS AND EQUIVALENT AOQs

This appendix addresses the matter of:

(1) The statement by Dodge (1943) in a footnote, "The solution given assumes correction or replacement of defective units. Where it is expedient to reject such units and not replace them, equations (19) to (22) inclusive, should be modified by replacing i by i - 1."

and,

(2) The statement by Dodge (1955b) in a footnote, "It can be shown that i should be increased by one in CSP-1 plans when defective units are removed but not replaced"—so that Procedure A1 (replacement) uses i versus i + 1 in Procedure A2 (non-replacement).

Some concern has been expressed that the above statements are contradictory—that they represent inconsistency with respect to (i - 1) versus (i + 1) in the application of CSP-1 under the two conditions of replacement and non-replacement. The following arguments demonstrate that the statements are **not** contradictory and that the applications are correct and consistent. The basis for the arguments is that in the earlier paper, Dodge (1943), the emphasis is on producing or obtaining the **equation** for the non-replacement case from the **equation** for the replacement case. For the latter paper, Dodge (1955b), the emphasis is on optional procedures with **equivalent AOQs** (i.e. AOQ **values**).

For each of the cases of (r)eplacement and (n)on-replacement, respectively,

$$AOQ_r(i;p,F) = p(1 - F) \tag{1}$$

and,

$$AOQ_n(i;p,F) = p(1 - F) / (1 - pF), \tag{2}$$

the former by Dodge (1943), and the latter by Case, et al (1973).

These, of course, are not equivalent for a given set of parameters, and the equations are not the same!

In fact, since $0 \le p \le 1$, $0 \le F \le 1$,

and hence $0 \le pF \le 1$, and $0 \le (1 - pF) \le 1$,

$$[p(1 - F)] / (1 - pF) \ge p(1 - F) ,$$

i.e. $AOQ_n(i) \ge AOQ_r(i)$ \hfill (3)

This is certainly intuitive, since AOQ_r is diluted (reduced) by the addition of corrected (or replacement by good) units—as opposed to AOQ_n not benefitting from such a dilution.

Hence, for a given set of parameters, including i, the AOQ for the non-replacement option will be larger, in general, than the AOQ for the replacement case. This will be noted later.

Equations:

It is possible to **obtain the equation** for $AOQ_n(i)$ **from** the equation for $AOQ_r(i)$, and vice versa with reverse substitution, by replacing i in $AOQ_r(i)$ with (i - 1), as follows:

For the CSP-1 procedure, $\quad F = f / (f + (1 - f)q^i),$ \hfill (4)

for parameters f, i, and q = (1 - p), as per Dodge (1943).

Hence, (1) becomes,

$$AOQ_r(i) = (p(1 - f)q^i) / (f + (1 - f)q^i), \qquad (5)$$

and, (2) becomes,

$$AOQ_n(i) = (p(1 - f)q^i) / (fq + (1 - f)q^i) \qquad (6)$$

Now, replacing i by $(i - 1)$ in the right hand side of (5) yields (note that no equals $(=)$ are used),

$$(p(1 - f)q^{i-1}) / (f + (1 - f)q^{i-1}),$$

$$[(p(1 - f)q^i)/q] / [(fq + (1 - f)q^i)/q],$$

$$(p(1 - f)q^i) / (fq + (1 - f)q^i),$$

which is the non-replacement $AOQ_n(i)$, i.e. **the non-replacement AOQ equation** with parameter i. But note that these equations (for given i) are **not equal**—as shown above,

$$AOQ_r(i) \le AOQ_n(i)$$

Hence, this procedure of replacing i (in $AOQ_r(i)$) by $(i - 1)$ has **produced the equation** for $AOQ_n(i)$. Note Dodge's reference to, "... the **equations** ... should be modified by replacing i by $(i - 1)$".

Equivalent AOQ:

Now, $\qquad AOQ_r(i; p, f) = (p(1 - f)q^i) / (f + (1 - f)q^i),$

and if we want to use the non-replacement procedure to get **this same value**; for given i, f, and p,

$$AOQ_n(j; p, f) = (p(1 - f)q^i) / (fq + (1 - f)q^i),$$

and multiplying by : $((1/q) / (1/q))$, i.e. both numerator and denominator by $(1/q)$,

$$AOQ_n(j; p, f) = (p(1 - f)q^{j-1}) / (f + (1 - f)q^{j-1}),$$

hence, for this **value** $(AOQ_n(j))$ to be **equal** to $AOQ_r(i)$, we must take $j - 1 = i$, or, $j = i + 1$

Example:

Let $p = .02$, $f = 1/2$, , and $i = 14$,

$$AOQ_r = [(.02)(.5)(.98)^{14}] / [.5 + (.5)(.98)^{14}]$$

$$= 0.0075364194 / 0.8768209$$

$$= 8.5951633 \times 10^{-3} \approx 0.008595$$

$$AOQ_n = [(.02)(.5)(.98)^{14}] / [(.5)(.98) + (.5)(.98)^{14}]$$

$$= 0.0075364194 / 0.8668209$$

$$= 8.6943206 \times 10^{-3} \approx 0.008694$$

and note that $AOQ_n > AOQ_r$

Now, let $p = .02$, $f = 1/2$, and $i = 15$,

$$AOQ_n = [(.02)(.5)(.98)^{15}] / [(.5)(.98) + (.5)(.98)^{15}]$$

$$= 0.007385691 / 0.8592845$$

$$= 8.5951633 \times 10^{-3} \approx 0.008595,$$

the same value as for AOQ_r with i = 14.

Hence, to obtain a non-replacement procedure that has equivalent AOQ to the replacement procedure (with same f) **increase** i (of the replacement procedure) by 1, i.e. use $i + 1$. This is what Dodge (1955b) was doing in proposing Procedures A1 and A2, i.e. the emphasis is on procedures having **equivalent AOQs**.

52

Tables I and II

ANSI/ASQC Standard S1-1987

TABLE I
MINIMUM CUMULATIVE SAMPLE SIZE TO INITIATE SKIP-LOT INSPECTION

NONCORMITIES OR NONCONFORMING ITEMS	AQL (Percent nonconforming* or nonconformities per hundred units)												
	0.1	0.15	0.25	0.40	0.65	1.0	1.5	2.5	4.0	6.5	10.0	15.0	25.0
0	2600	1740	1040	650	400	260	174	104	65	40	26	17	10
1	4250	2840	1700	1070	654	425	284	170	107	65	43	28	17
2	5740	3830	2300	1440	883	574	383	230	144	88	57	38	23
3	7140	4760	2860	1790	1098	714	476	286	179	110	71	48	29
4	8490	5660	3400	2120	1306	849	566	340	212	131	85	57	34
5	9800	6530	3920	2450	1508	980	653	392	245	151	98	65	39
6	11090	7390	4440	2770	1706	1109	739	444	277	171	111	74	44
7	12360	8240	4940	3090	1902	1236	824	494	309	190	124	82	49
8	13610	9070	5440	3400	2094	1361	907	544	340	209	136	91	54
9	14850	9900	5940	3710	2285	1485	990	594	371	229	149	99	59
10	16080	10720	6430	4020	2474	1608	1072	643	402	247	161	107	64
11	17290	11530	6920	4320	2660	1729	1153	692	432	266	173	115	69
12	18500	12330	7400	4630	2846	1850	1233	740	463	285	185	123	74
13	19700	13130	7880	4930	3031	1970	1313	788	493	303	197	131	79
14	20890	13930	8360	5220	3214	2089	1393	836	522	321	209	139	84
15	22080	14720	8830	5520	3397	2208	1472	883	552	340	221	147	88
16	23260	15500	9300	5820	3578	2326	1550	930	582	358	233	155	93
17	24430	16290	9770	6110	3758	2443	1629	977	611	376	244	163	98
18	25600	17070	10240	6400	3938	2560	1707	1024	640	394	256	171	102
19	26760	17840	10700	6690	4117	2676	1784	1070	669	412	268	178	107
20	27930	18620	11170	6980	4297	2793	1862	1117	698	430	279	186	112
†n=	1170	780	470	290	180	117	78	47	29	18	12	8	5

* Percent nonconforming applies only to AQL values of 10 or less.

† For each additional nonconformity or nonconforming item add n to the minimum cumulative sample size for 20 nonconformities. For example, at an AQL of 1.0%, 22 nonconformities are noted. The minimum cumulative sample size is calculated as follows:

$$2 \times 117 + 2793 = 3027$$

TABLE II
ACCEPTANCE NUMBERS TO INITIATE OR CONTINUE SKIP-LOT INSPECTION
(INDIVIDUAL LOT CRITERION)

SAMPLE SIZE	AQL (Percent nonconforming* or nonconformities per hundred units)												
	0.1	0.15	0.25	0.4	0.65	1.0	1.5	2.5	4.0	6.5	10.0	15.0	25.0
2	-	-	-	-	-	-	-	-	-	0	-	0	1
3	-	-	-	-	-	-	-	-	0	-	0	1	1
5	-	-	-	-	-	-	-	0	-	0	1	1	2
8	-	-	-	-	-	-	0	-	0	1	1	2	3
13	-	-	-	-	-	0	-	0	1	1	2	3	5
20	-	-	-	-	0	-	0	1	1	2	3	5	7
32	-	-	-	0	-	0	1	1	2	3	5	7	11
50	-	-	0	-	0	1	1	2	3	5	7	11	17
80	-	0	-	0	1	1	2	3	5	7	11	17	-
125	0	-	0	1	1	2	3	5	7	11	17	-	-
200	-	0	1	1	2	3	5	7	11	17	-	-	-
315	0	1	1	2	3	5	7	11	17	-	-	-	-
500	1	1	2	3	5	7	11	17	-	-	-	-	-
800	1	2	3	5	7	11	17	-	-	-	-	-	-
1250	2	3	5	7	11	17	-	-	-	-	-	-	-
2000	3	5	7	11	17	-	-	-	-	-	-	-	-

* Percent nonconforming applies only to AQL values of 10 or less.

REFERENCES

Anon. 1970. Dependent State Sampling Plans, *Research Report QR-TR-70-2*, Technology and Programs Division, Product Assurance and Test Management Office, U.S. Army Missile Command, Redstone Arsenal, Alabama.

————. 1982. Chain Sampling, *Encyclopedia of Statistical Science*, Vol 2, Kotz, S., Johnson N. L., and Read, C. B. (Editors), John Wiley and Sons, Inc., New York, pp. 402–403.

————. 1985. Skip-Lot Sampling, *Quadripartite Advisory Publication 28*, American-British-Canadian-Australian Armies Standardization Program, November 1985.

Anscombe, F. J., H. J. Godwin, and R. L. Plackett. 1947. Methods of Deferred Sentencing in Testing the Fraction Defective of a Continuous Output, *Supplement to the Journal of the Royal Statistical Society*, Vol. IX, Nos. 1-2, pp. 198–217.

ANSI/ASQC Standard A2-1987. 1987. *Terms, Symbols, and Definitions for Acceptance Sampling*, American Society for Quality Control, Milwaukee, Wisc.

ANSI/ASQC Standard S1-1987. 1987. *An Attribute Skip-Lot Sampling Program*, American Society for Quality Control, Milwaukee, Wisc.

ANSI/ASQC Standard Z1.4. 1981. *Sampling Procedures and Tables for Inspection by Attributes*, American Society for Quality Control, Milwaukee, Wisc.

Asanaiye, P. A. 1983. Chain-Deferred Inspection Plans, *Journal of Applied Statistics*, 32, No. 1, pp. 19–24.

————. 1985. Multiple Chain-Deferred Inspection Plans and Their Compatibility with the Multiple Plans in MIL-STD-105D and Equivalent Schemes, *Journal of Applied Statistics*, 12, No. 1, pp. 71–81.

ASQC Statistics Division. 1983. *Glossary and Tables for Statistical Quality Control*, ASQC Quality Press, Milwaukee, Wisc.

Banks, J. 1989. *Principles of Quality Control*, John Wiley and Sons, Inc., New York, N.Y., pp. 441–458.

Beattie, D. W. 1962. A Continuous Acceptance Sampling Procedure Based Upon a Cumulative Sum Chart for the Number of Defectives, *Applied Statistics*, Vol. 11, No. 3, November 1962, pp. 137–147.

Bennett, G. K., and C. J. Callejas. 1980. The Economic Design of Skip Lot Sampling Plans, *AIIE Proceedings—Spring Annual Conference*, May 13, 1980, pp. 349–355.

Bloom, A. G. 1968. Ratio/Skip-Lot Sampling, A New Approach to Government Product Verification, *Annual Technical Conference Transactions*, American Society for Quality Control, Milwaukee, Wisc., pp. 53–59.

Breeze, J. D., and J. J. Heldt. 1982. Selecting Sampling Procedures Can Be Fun, *36th Annual Technical Conference Transactions*, ASQC, Milwaukee, Wisc., pp. 709–716.

Brugger, R. M. 1974. Skip-Lot Procedure Using the Simplified Markov Chain Method, *Proceedings of the 19th Conference on the Design of Experiments in Army Research, Development and Testing, Part 2, ARO Report 75-2*, October 1974, pp. 647–656.

————. 1975. A Simplification of Skip-Lot Procedure Formulation, *Journal of Quality Technology*, Vol. 7, No. 4, October 1975, pp. 165–167

Callejas, C. J. 1976. The Economic Design of a System of Skip Lot Sampling Plans, Master's Thesis, University of South Florida, December 1976.

Carr, W. E. 1982. Sampling Plan Adjustment for Inspection Error and Skip-Lot Plan, *Journal of Quality Technology*, Vol. 14, No. 3, July 1982, pp. 105–116.

Case, K. E., E. G. Bennett, and J. W. Schmidt. 1973. The Dodge CSP-1 Continuous Sampling Plan Under Inspection Error, *AIIE Transactions*, Vol. 5, No. 3, September 1973, pp. 193–202.

Clark, C. R. 1960. OC Curves for ChSP-1 Chain Sampling Plans, *Industrial Quality Control*, Vol. 17, No. 4, October 1960, pp. 10–12.

Clark, D. L. 1977. U.S. Army Applications of Chain Sampling, *Interim Note No. R-62*, Reliability, Availability and Maintainability Division, U.S. Army Material Systems Analysis Activity, Aberdeen Proving Ground, Maryland, December 1977.

Cone, A. F., and H. F. Dodge. 1964. A Cumulative-Results Plan for Small-Sample Inspection, *Industrial Quality Control*, Vol. 21, No. 1, July 1964, pp. 4–9.

Cox, C. C. 1982. Skip-Lot Sampling Plan, *Quality*, Hitchcock Publishers, Wheaton, Ill., August 1982, pp. 26–27.

Derman, C. S., S. Littauer, and H. Solomon. 1957. Tightened Multi-Level Continuous Sampling Plans, *The Annals of Mathematical Statistics*, Vol. 28, No. 2, June 1957, pp. 395–404. (Also published in *Technical Report No. 28*, Applied Mathematics and Statistics Laboratory, Stanford University, Stanford, Calif., June 20, 1956.)

Dodge, H. F. 1943. A Sampling Inspection Plan for Continuous Production, *The Annals of Mathematical Statistics*, Vol. 14, No. 3, September 1943, pp. 264–279 and *Transactions of the ASME*, Vol. 66, No. 2, February 1944, pp. 127–133; and also reprinted in the *Journal of Quality Technology*, Vol. 9, No. 3, pp. 104–119, July 1977.

———. 1955a. Chain Sampling Inspection Plan, *Industrial Quality Control*, Vol. 11, No. 4, January 1955, pp. 10–13.

———. 1955b. Skip-lot Sampling Plan, *Industrial Quality Control*, Vol. 11, No. 5, February 1955, pp. 3–5.

———. 1958. Chain Sampling Plans—ChSP-2 and ChSP-3, unpublished memorandum, Bell Telephone Laboratories, Murray Hill, N. J., October 1958.

———. 1970. Notes on the Evolution of Acceptance Sampling Plans, Part IV, *Journal of Quality Technology*, Vol. 2, No. 1, January 1970, pp. 1–8.

Dodge, H. F., and R. L. Perry. 1971. A System of Skip-Lot Plans for Lot by Lot Inspection, *Annual Technical Conference Transactions*, American Society for Quality Control, Milwaukee, Wisc., pp. 469–477.

Dodge, H. F., and H. G. Romig. 1959. *Sampling Inspection Tables—Single and Double Sampling*, John Wiley & Sons, New York, 2nd edition, 1959.

Dodge, H. F., and K. S. Stephens. 1964. A General Family of Chain Sampling Inspection Plans, *Technical Report No. N-20*, Rutgers-The State University of New Jersey, Statistics Center, New Brunswick, N. J., December 1964.

———. 1965. Some New Chain Sampling Inspection Plans, *Annual Technical Conference Transactions*, American Society for Quality Control, Milwaukee, Wisc., May 1965, pp. 8–17.

———. 1966. Some New Chain Sampling Inspection Plans, *Industrial Quality Control*, Vol. 23, No. 2, August 1966, pp. 61–67.

Duncan, A. J. 1986. *Quality Control and Industrial Statistics*, 5th edition, Richard D. Irwin, Homewood, Ill.

Endres, A. 1967a. A Model for Obtaining the Operating Characteristics of a Skip Lot Sampling Procedure, U.S. Army Ammunition Procurement and Supply Agency, Quality Assurance Directorate, Quality Evaluation Division, Concepts Branch, Joliet, Ill.

———. 1967b. Derivation of the Operating Characteristic Curve Formula of Project SKIP, *QEM 21-230-3*, U.S. Army Ammunition Procurement and Supply Agency, Quality Assurance Directorate, Quality Evaluation Division, Concepts Branch, Joliet, Ill., July 1967.

———. 1967c. The Derivation of an Expected Ratio Used in Determining the Operating Characteristics of Project SKIP, *QEM 21-240-6*, U.S. Army Ammunition Procurement and Supply Agency, Quality Assurance Directorate, Quality Evaluation Division, Concepts Branch, Joliet, Ill., July 1967.

———. 1967d. The Expected Reduction in Ballistic Testing Through Project SKIP, *QEM 21-240-7*, U.S. Army Ammunition Procurement and Supply Agency, Quality Assurance Directorate, Quality Evaluation Division, Concepts Branch, Joliet, Ill., October 1967.

———. 1968. The Derivation of the Operating Characteristic Curve of a Skip-Lot Sampling Plan, *Proceedings of the 13th Conference on the Design of Experiments in Army Research, Development and Testing, ARO Report 68-2*, 1–3 November 1968.

Enkawa, T. 1984. Cu-Sum Sampling Plan for Continuous Production and Its Application, *Reports of Statistical Applications Research*, JUSE, Vol. 31, No. 1, March 1984, pp. 19–33.

Ewan, W. D., and K. W. Kemp. 1960. Sampling Inspection of Continuous Processes With No Autocorrelation Between Successive Results, *Biometrika*, Vol. 47, December 1960, pp. 363–380.

Frishman, F. 1954. Standard Sampling Plans for Torpedo Group Proving, *NAVORD Instruction 8510.40*, Navy Department, Bureau of Ordnance, November 12, 1954.

———. 1960. An Extended Chain Sampling Inspection Plan, *Industrial Quality Control*, Vol. 17, No. 1, July 1960, pp. 10–12.

Harishchandra, K., and T. Srivenkataramana. 1982. Link Sampling for Attributes, *Communication in Statistics—Theory and Methods*, Vol 11, pp. 1855–1868.

Heldt, J. J. 1981. Continuous Sampling Plans and Skip Lot Application, *Quality*, Hitchcock Publishers, Wheaton, Ill., June 1981, pp. 65–66.

Hill, I. D., G. Horsnell, and B. T. Warner. 1959. Deferred Sentencing Schemes, *Applied Statistics*, Vol. 8. pp. 76–91.

Hopkins, C. L., R. L. Nutt, and R. Heathcock. 1972. Chain Sampling, *Technical Report QP-TR-72-1*, Advanced Techniques Branch, Plans and Programs Analysis Division, Directorate for Product Assurance, U.S. Army Missile Command, Redstone Arsenal, Alabama, 1 August 1972.

Hsu, J. I. S. 1977. A Cost Model for Skip-Lot Destructive Sampling, *IEEE Transactions on Reliability*, Vol. R-26, No. 1, April 1977, pp. 70–72.

———. 1980. An Economic Design of Skip-Lot Sampling Plans, *Journal of Quality Technology*, Vol. 12, No. 3, July 1980, pp. 144–149.

Jafri, S. Q. 1988. Skip Parts Continuous Sampling Plans for Kanban (Just-In-Time) Environment, With Three Dimension Decision Criteria, *Proceedings of the 32nd EOQC Annual Conference*, Moscow, USSR, 13–17 June 1988, pp. 120–126.

Johnson, N. L., S. Kotz, and R. N. Rodriguez. 1986. Statistical Effects of Imperfect Inspection Sampling: II. Double Sampling and Link Sampling, *Journal of Quality Technology*, Vol. 18, No. 2, April 1986, pp. 116–138.

Juran, J. M. 1988. (Editor-in-Chief), *Quality Control Handbook*, 4th edition, McGraw-Hill Book Company, New York.

Kemp, K. W. 1962. The Use of Cumulative Sums for Sampling Inspection Schemes, *Applied Statistics*, Vol. 11, No. 1, March 1962, pp. 16–31.

Lenz, H.-J., and U. Rendtel. 1984. Performance Evaluation of the MIL-STD-105D, Skip-Lot Sampling Plans, and Bayesian Single Sampling Plans, contributing chapter to: *Frontiers in Statistical Quality Control*, Second Edition, Edited by H.-J. Lenz, G. B. Wetherill, and P.-Th Wilrich, Physica-Verlag, Würzburg, Germany.

Lieberman, G. J., and H. Solomon. 1955. Multi-Level Continuous Sampling Plans, *The Annals of Mathematical Statistics*, Vol. 26, No. 4, December 1955, pp. 686–704. (Also published in *Technical Report No. 17*, Applied Mathematics and Statistics Laboratory, Stanford University, Stanford, California, September 1954.)

Liebesman, B. S., and F. C. Hawley. 1984. Small Acceptance Number Plans for Use in Military Standard 105D, *Journal of Quality Technology*, Vol. 16, No. 4, October 1984, pp. 219–231.

Liebesman, B. S., and B. Saperstein. 1983. A Proposed Attribute Skip-Lot Sampling Program, *Journal of Quality Technology*, Vol. 15, No. 3, July 1983, pp. 130–140.

MIL-STD-105E. 1989. *Sampling Procedures and Tables for Inspection by Attributes*, United States Department of Defense, Washington, D.C., 10 May 1989.

Mundel, A. B. 1989–90. STANDARDS COLUMN—Skip-Lot Sampling Standards, *Quality Engineering*, Vol. 2, No. 2, pp. 163–172.

———. 1990. STANDARDS COLUMN—Skip-Lot Sampling, *Quality Engineering*, Vol. 2, No. 4, pp. 487–495.

Osborne, J. 1990–91. ''Skip-To-Stock'' Plan: Expedient and Cost-Effective, *Quality Engineering*, Vol. 3, No. 1, pp. 1–8.

Page, E. S. 1955. Continuous Inspection Schemes, *Biometrika*, Vol. 41, Nos. 1–2, June 1955, pp. 100–115.

Parker, R. D., and L. Kessler. 1981. A Modified Skip-Lot Sampling Plan, *Journal of Quality Technology*, Vol. 13, No. 1, January 1981, pp. 31–35.

Perry, R. L. 1970. A System of Skip-Lot Sampling Plans for Lot Inspection, unpublished Ph.D. Dissertation, Rutgers-The State University of New Jersey, Statistics Center, New Brunswick, N.J., October 1970.

———. 1973a. Skip-Lot Sampling Plans, *Journal of Quality Technology*, Vol. 5, No. 3, July 1973, pp. 123–130.

———. 1973b. Two-Level Skip-Lot Sampling Plans—Operating Characteristic Properties, *Journal of Quality Technology*, Vol. 5, No. 4, October 1973, pp. 160–166.

Pesotchinsky, L. 1987. Plans for Very Low Fraction Nonconforming, *Journal of Quality Technology*, Vol. 19, No. 4, October 1987, pp. 191–196.

Phelps, R. I. 1982. Skip-Lot Destructive Sampling with Bayesian Inference, *IEEE Transactions on Reliability*, Vol. R-31, No. 2, June 1982, pp. 191–193.

Prairie, R. R., and W. J. Zimmer. 1973. Graphs, Tables and Discussion to Aid in the Design and Evaluation of an Acceptance Sampling Procedure Based on Cumulative Sums, *Journal of Quality Technology*, Vol. 5, No. 2, April 1973, pp. 58–66.

Reetz D. 1984. Optimal Skip-Lot Single Sampling Plans for Markov Chains—Theoretical Foundation, contributing chapter to: *Frontiers in Statistical Quality Control*, Second Edition, Edited by H.-J. Lenz, G. B. Wetherill, and P.-Th Wilrich, Physica-Verlag, Würzburg, Germany.

Reimann, A. 1982. Skip-Lot is not Optimal in the Case of Independent Fractions Defective—Comments on a Paper by Professor Hsu, *Journal of Quality Technology*, Vol. 14, No. 4, October 1982, pp. 211–214.

Schilling, E. G. 1982. *Acceptance Sampling in Quality Control*, Marcel Dekker, New York.

Soundararajan, V. 1978a. Procedures and Tables for Construction and Selection of Chain Sampling Plans (ChSP-1), Part 1, *Journal of Quality Technology*, Vol. 10, No. 2, April 1978, pp. 56–60.

———. 1978b. Procedures and Tables for Construction and Selection of Chain Sampling Plans (ChSP-1), Part 2, *Journal of Quality Technology*, Vol. 10, No. 3, July 1978, pp. 99–103.

Soundararajan, V., and K. Govindaraju. 1982. A Note on Designing Chain Sampling Plans (ChSP-1), *QR Journal*, Indian Association of Quality and Reliability, Vol. 9, No. 3, September 1982, pp. 121–123.

———. 1983. Construction and Selection of Chain Sampling Plans ChSP-(0,1), *Journal of Quality Technology*, Vol. 15, No. 4, October 1983, pp. 180–185.

Soundararajan, V., and C. Raju. 1984. Procedures and Tables for Construction of Chain Sampling Plans ChSP-4 (c_1,c_2), *IAPQR Transactions*, Vol. 9, No. 2, pp. 41–65.

Stephens, K. S. 1966. A Class of Cumulative-Results Sampling Inspection Plans and Their Evaluation, an unpublished Ph.D. Dissertation, Rutgers-The State University of New Jersey, Statistics Center, New Brunswick, N. J., May 1966.

———. 1979. How to Perform Continuous, Skip-Lot and Chain Sampling, *Annual Technical Conference Transactions*, American Society for Quality Control, Milwaukee, Wisc., pp. 207–208.

———. 1980. How to Perform Skip-Lot and Chain Sampling, *Annual Technical Conference Transactions*, American Society for Quality Control, Milwaukee, Wisc., pp. 53–55.

———. 1995. How to Perform Continuous Sampling, *The ASQC Basic References in Quality Control: Statistical Techniques*, Vol. 2, American Society for Quality Control, Milwaukee, Wisc.

Stephens, K. S., and H. F. Dodge. 1965a. Chain Sampling Inspection Plans—ChSP-0,2 and ChSP-1,2, *Technical Report No. N-21*, Rutgers-The State University of New Jersey, Statistics Center, New Brunswick, N. J., May 1965.

———. 1965b. Chain Sampling Inspection Plans—ChSP-0,3 and ChSP-1,3, *Technical Report No. N-22*, Rutgers-The State University of New Jersey, Statistics Center, New Brunswick, N. J., December 1965.

———. 1966a. Chain Sampling Inspection Plans—ChSP-0,4 and ChSP-1,4, *Technical Report No. N-24*, Rutgers-The State University of New Jersey, Statistics Center, New Brunswick, N. J., February 1966.

————. 1966b. Two-Stage Chain Sampling Inspection Plans with Different Sample Sizes in the Two Stages, *Technical Report No. N-23*, Rutgers-The State University of New Jersey, Statistics Center, New Brunswick, N. J., April 1966.

————. 1966c. Comparison of Chain Sampling Plans With Single and Double Sampling Plans, *Transactions of the 18th Annual All Day Conference on Quality Control*, ASQC-Metropolitan Section and Rutgers-The State University of New Jersey, Statistics Center, New Brunswick, N. J., September 10, 1966, pp. 126–131.

————. 1967. Evaluation of Response Characteristics of Chain Sampling Inspection Plans, *Technical Report No. N-25*, Rutgers-The State University of New Jersey, Statistics Center, New Brunswick, N. J., March 1967.

————. 1974. An Application of Markov Chains for the Evaluation of the Operating Characteristics of Chain Sampling Inspection Plans, *The QR Journal*, Indian Association for Quality and Reliability, Bangalore, September 1974, pp. 131–138.

————. 1976a. Comparison of Chain Sampling Plans With Single and Double Sampling Plans, *Journal of Quality Technology*, Vol. 8, No. 1, January 1976, pp. 24–33.

————. 1976b. Two-Stage Chain Sampling Inspection Plans with Different Sample Sizes in the Two Stages, *Journal of Quality Technology*, Vol. 8, No. 4, October 1976, pp. 209–224.

Stephens, K. S., and K. E. Larson. 1967. An Evaluation of the MIL-STD-105D System of Sampling Plans, *Industrial Quality Control*, Vol. 23, No. 7, January 1967.

Stine, C. E. 1974. Project SKIP—An Idea Takes Hold, *The Army Logistician*, Vol. 6, No. 3, May–June 1974, pp. 30–31.

Taylor, W. A. 1993. Quick Switching Systems for Acceptance Sampling, *1993 ASQC Quality Congress Transactions—Boston*, May 1993, pp. 496–502, ASQC, Milwaukee, Wisc.

Taylor, W. A. 1992. *Guide to Acceptance Sampling*, Taylor Enterprises, Lindenhurst, Ill.

Wadsworth, H. M., K. S. Stephens, and A. B. Godfrey. 1986. *Modern Methods for Quality Control and Improvement*, John Wiley and Sons, New York.

Wortham, A. W., and J. M. Moog. 1970. Dependent Stage Sampling Inspection, *The International Journal of Production Research*, Vol. 8, No. 4, pp. 385–395.

Zwickl, R. D. 1965. Chain Sampling for Average Quality Protection, Term Paper, Rutgers-The State University of New Jersey, Statistics Center, New Brunswick, N. J., January 4, 1965.

INDEX